An Guide

Trees, and Herbs in Your Path

Maggie Herskovits

MICROCOSM PUBLISHING
Portland, OR | Cleveland, OH

An Urban Field Guide to the Plants, Trees, and Herbs in Your Path

Part of the DIY Series

© Maggie Herskovits 2025
This edition © Microcosm Publishing, 2025
First Edition, 3,000 copies, first published Feb 2025

ISBN 9781648413292
This is Microcosm #844
Designed by Joe Biel
Edited by Kandi Zeller
Illustrated by Maggie Herskovits
For a catalog, write or visit:
Microcosm Publishing
2752 N Williams Ave.
Portland, OR 97227
(503)799-2698
Microcosm.Pub

Did you know that you can buy our books directly from us at sliding scale rates? Support a small, independent publisher and pay less than Amazon's price at **www.Microcosm.Pub.**

All the news that's fit to print at www.Microcosm.Pub/Newsletter

Get more copies of this book at www.Microcosm.Pub/FieldGuide

Find more work by Maggie Herskovits at www.Microcosm.Pub/MaggieHerskovits

To join the ranks of high-class stores that feature Microcosm titles, talk to your rep: In the U.S. **COMO** (Atlantic), **ABRAHAM** (Midwest), **BOB BARNETT** (Texas, Oklahoma, Arkansas, Louisiana), **IMPRINT** (Pacific), **TURNAROUND** (UK), **UTP/MANDA** (Canada), **NEWSOUTH** (Australia/New Zealand), **Observatoire** (Africa, Europe), **IPR** (Middle East), **Yvonne Chau** (Southeast Asia), **HarperCollins** (India), **Everest/B.K. Agency** (China), **Tim Burland** (Japan/Korea), and **FAIRE** in the gift trade.

Global labor conditions are bad, and our roots in industrial Cleveland in the 70s and 80s made us appreciate the need to treat workers right. Therefore, our books are MADE IN THE USA and printed on post-consumer paper.

MICROCOSM PUBLISHING is Portland's most diversified publishing house and distributor, with a focus on the colorful, authentic, and empowering. Our books and zines have put your power in your hands since 1996, equipping readers to make positive changes in their lives and in the world around them. Microcosm emphasizes skill-building, showing hidden histories, and fostering creativity through challenging conventional publishing wisdom with books and bookettes about DIY skills, food, bicycling, gender, self-care, and social justice. What was once a distro and record label started by Joe Biel in a drafty bedroom was determined to be *Publishers Weekly*'s fastest-growing publisher of 2022 and #3 in 2023, and is now among the oldest independent publishing houses in Portland, OR, and Cleveland, OH. Biel is also the winner of PubWest's Innovator Award in 2024. We are a politically moderate, centrist publisher in a world that has inched to the right for the past 80 years.

To my grandmothers: Clarice Hoffer and Agnes Herskovits
Thank you to all the people and plants that made this book possible

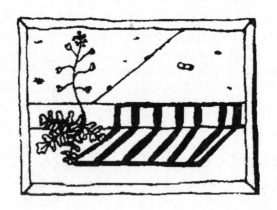

"I really believe that to stay home, to learn the names of things, to realize who we live among... The notion that we can extend our sense of community, our idea of community, to include all life forms—plants, animals, rocks, rivers and human beings—then I believe a politics of place emerges where we are deeply accountable to our communities, to our neighborhoods, to our home. Otherwise, who is there to chart the changes?"

—Terry Tempest Williams, "The Politics of Place: An Interview with Terry Tempest Williams"

Table of Contents

INTRODUCTION • 8

HERBACEOUS PLANTS • 14

    Asiatic Dayflower • 15
    Black Nightshade • 18
    Bladder Campion • 21
    Broadleaf Plantain • 24
    Burdock • 28
    Canada Goldenrod • 31
    Chickweed • 34
    Chicory • 37
    Cleavers • 40
    Coltsfoot • 43
    Common Blue Violet • 46
    Crown Vetch • 49
    Curly Dock • 52

    Daisy Fleabane • 56
    Dandelion • 59
    Evening Primrose • 62
    Galinsoga • 65
    Garlic Mustard • 68
    Green Foxtail • 71
    Ground Ivy • 74
    Hairy Bittercress • 77
    Hedge Bindweed • 80
    Horseweed • 83

    Japanese Knotweed • 87
    Jewelweed • 91
    Lambsquarters • 94
    Mallow • 97
    Milkweed • 100
    Mugwort • 103

Mullein • 107
Pennsylvania Smartweed • 110
Pineapple Weed • 113
Pokeweed • 116
Prickly Lettuce • 120
Purslane • 123
Quackgrass • 126
Ragweed • 129
Red Clover • 132
Redroot Pigweed • 135
Reed • 138
Shepherd's Purse • 141
Smooth Crabgrass • 144
Spotted Spurge • 147
St. John's Wort • 150
Stinging Nettles • 153
Virginia Pepperweed • 156
Wild Carrot • 159
Yellow Rocket • 162
Yellow Woodsorrel • 165

WOODY PLANTS • 168
Black Cherry • 170
Empress Tree • 173
Hackberry • 176
Norway Maple • 181
Quaking Aspen • 184
Staghorn Sumac • 188
Tree-of-Heaven • 192
Virginia Creeper • 196
White Mulberry • 199

RESOURCES AND FURTHER READING • 203
GLOSSARY OF TERMS • 206

Introduction

I got to know many of the plants in this book intimately during my time spent as a gardener in a New York City park. I first learned their names and how to identify them in order to rip them from the soil. In our clean garden plan, they were considered invaders. We used to joke as gardeners that weeds were our job security, for as long as there were weeds we had a job to do. And there were always weeds.

And then one day, two years in, I saw things differently. What I saw was a lone lambsquarter growing discreetly within a patch of *Persicaria virginiana*. A seed had taken advantage of a small opening that allowed just enough light to sprout. The plant then took on a very slender form to fit in this hiding place and stopped growing just when it reached the height of the surrounding Persicaria. Genius. Buds were bulging, ready to blossom, aliveness was gushing. You could not see it from the edge of the garden bed, and so it was not interrupting the integrity of the planting plan. My first act of rebellion was to let it live.

Up until that moment I prided myself in being able to find the most hidden weeds, remove the subversive beings and keep the garden clean. From that day forward, I joined in on their subversion. I still found them alright and when I did, I knelt close and marveled at just how cunning these plants were.

This gardener's existential crisis brought me to know these plants beyond the label of weed. I got to know them for who they are, up close and personal. They learned about me, and we shared lessons and stories of life. Come take a walk with me on the pathway to relationship with spontaneous urban plants. You know, the plants on the block that seem to come from nowhere and grow from thin air. But everything is connected and nothing comes from nothing. Seeds are spread by birds, wind, and you. They land in cracks of concrete, vacant lots, and other city habitats. The newly sprouted plants grow to form seeds of their own and thus continue the cycle.

• • •

This field guide is filled with stories of plants that inhabit the cities of North America (Turtle Island). Some were present pre-colonization, some were introduced intentionally, and still others traveled here using us humans and other animals as unwitting carriers—mixed into boat ballast, hidden in bags of grain, tucked into packing materials, or stuck in cow hooves, just a few examples of the endless possibilities of plant travel.

No matter how or when they got to this land, the plants that thrive in the city do so because of their tolerance to all of the challenging factors that make up the urban environment. Even though the city is a relatively new place here on Earth, and no plant grew up or evolved here, the adaptations plants developed to survive are well suited to the urban environment. The plants that grace our urban home developed their survival traits over countless generations to become a perfect fit for the place where they were born. Plants that thrive in the urban environment have evolved in places of Mother Earth's constant shifts and edges. As Richard Mabey writes:

"A good proportion . . . made a life in the planet's most restless places. They'd evolved on tide pounded beaches and the precarious slopes of volcanoes, in the flood zones at the edges of rivers and the muddy wallows made by wild grazing animals, in scree and shingle and glacial moraines."

They do well here because the urban ecosystem is also a restless place, a fact illuminated in its Latinate name: ruderal ecosystem. Ruderal comes from the Latin word *rudus*, meaning "rubble." Here in the rubble lives are lived—plants, insects, birds, raccoons, squirrels, dogs, and more—all interacting with each other and the environment, an active ecosystem. That's right, city plants and animals (yes, that includes us humans) are among the living organisms that interact within the ruderal ecosystem.

For the most part, the urban ecosystem is human-made and human-run, but our structures still exist upon and within the complex systems of Mother Earth. Our cities are subject to her weather as well as her rare, yet formidable, natural disasters, but the everyday actions of humans create the conditions that form the environment of the urban ecosystem.

In ecosystems governed by Mother Earth, there is a process called ecological succession, the process by which natural communities replace, or succeed, each other over time. The last stage of succession, a climax community, is a high biomass, high complexity ecosystem like a mature forest. This stage remains until a disturbance (fire, flood, hurricane, earthquake, lava flow, etc.) renews the process by erasing what is there, leaving a blank slate for something new to start. Like the change of seasons, ecological succession moves through this cycle in a circular fashion, over time, again and again.

The depleted land created by the disturbance is then colonized by plants known as pioneer species. They have adapted to be quick to sprout, quick to grow, and quick to set seed. Some annual plants, like green foxtail and chickweed, can have up to five generations in one growing season! Pioneer species generally love full sun and tolerate a range of challenging growing conditions: heat, drought, compaction, low-nutrient soil, air pollution, salt, and a wide range of soil pH. The pioneer species slowly attract wildlife and make the land more habitable for the larger, more specialized plants that will eventually shade them out, creating a new ecosystem and continuing the slow march toward the climax community.

In the urban ecosystem, the process of Mother Earth's ecological succession is thwarted by humans. The constant construction, concrete, and change keeps the urban ecosystem in a sort of arrested development, suppressed by its defining feature: disturbance. City land resembles the native habitats of the plants known as pioneer species in that it shares the challenging growing conditions that other plants cannot tolerate. So it is the pioneer species that greet us every spring in the oft-disturbed urban home we share. With little competition from larger plants that may succeed them, their place each year is guaranteed.

• • •

As of this writing, more than 80 percent of Americans live in cities, and according to the U.N., 60 percent of the world's population will live in cities by 2050. The urban ecosystem is growing rapidly at the expense of other ecosystems whose communities are being disrupted. While we grieve the loss of landscapes and species, we can find hope in these city plants. They provide needed diversity as a source of food and shelter for bees, birds, and butterflies. And, just as important, they provide medicine for our souls as a source of contact to the natural world when it may seem like nature is far, far away.

In the pages to come, each plant profile includes **key features** that will help you identify the plant in the field. Also included is information on

- geography
- historical uses
- fun facts

The plants are listed alphabetically by common name. Entries include the plants' botanical Latin names. Common names are made by the people in the language of the people, so you may know plants by different names. All the better!

This book is not a guidebook for foraging nor is it an herbal medicine guide. Please consult professional foragers and herbalists (there are many!) if you are interested in having a deeper relationship with the plants outlined in this book. Please do not harvest any plant unless you are 100 percent certain of its identity. This book is simply the meet and greet.

If you are ready to harvest after reading this book, here are some tips on foraging:

- Always ask first.
- Take no more than 10 percent.

- Take only what you need.
- Make sure no chemicals have been sprayed. If lambsquarter or dandelion are present, it could be a sign of safety (they are very susceptible to herbicides).
- A general rule is to harvest at least twenty feet away from a busy roadway to avoid contaminants.
- Be wary of compounds in the soil. Plants can take up toxins in their leaves and then pass those on to you.

This book helps us connect to our urban place, name the plants, recognize their survival skills, and admire their brilliance. When we learn the names of beings, we begin to show them love. Robin Wall Kimmerer writes, "It is a sign of respect to call a being by its name, and a sign of disrespect to ignore it. Words and names are the ways we humans build relationship, not only with each other, but also with plants." Erich Fromm helps to close the circle, connecting naming, respect, and love. He writes, "The active character of love becomes evident in the fact that it always implies certain basic elements, common to all forms of love. These are care, responsibility, respect and knowledge." When we are surrounded by beings we love, this experience brings us to love our place, the land, and gives us the capacity and desire to treat our home with care.

And finally, to know the plants in your path is to forget loneliness. Present at just about every time of the year, they are friends to sit with and share quiet existence.

Herbaceous Plants

Here we meet the urban plants that have only soft, green tissue growth above ground; there is no wood to be seen. Beyond that common trait, there are many things that make them shine as individuals. As if they didn't give the world enough with their beauty, most of the plants listed in this section have a long history of use as medicinal and/or edible herbs. Whatever place they originated from and have since traveled to, they bring stories and histories of the people with whom they lived. Plants have so much to teach us, even about our own species' history. Wisdom that has been passed down for generations is still accessible today. These plants can heal us in so many ways.

Many urban plants are healers of the land as well. Some repair soil by adding nutrients when there are none or breaking up compacted soil. Others do so by removing toxic minerals and compounds that poison the soil. They store the toxins in their leaves and stems, leaving behind a soil cleansed in a process called phytoremediation.

Finally, in their urban home, herbaceous plants add beauty, softness, and a splash of color to the concrete jungle. They are neighbors that share the joys and sorrows, highs and lows of living in a big city. Come commiserate, they beckon.

Asiatic Dayflower

Commelina communis

An annual of the Commelinaceae family, also known as the Spiderwort family, the Asiatic dayflower is related to the common houseplants of the same genus. In temperate climates, such as Northeastern US (where I live), we are most familiar with plants of this family as houseplants because, in the wild, this family is mainly found in tropical and subtropical regions.

Habitat: Enjoys moist, shady areas and can be found in parks, along building foundations, sidewalk cracks, tree pits, and along waterways.

Came here from Asia. Maybe in packing materials? Asiatic dayflower has traveled as far West as Nebraska and was first reported growing there in 1905.

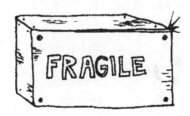

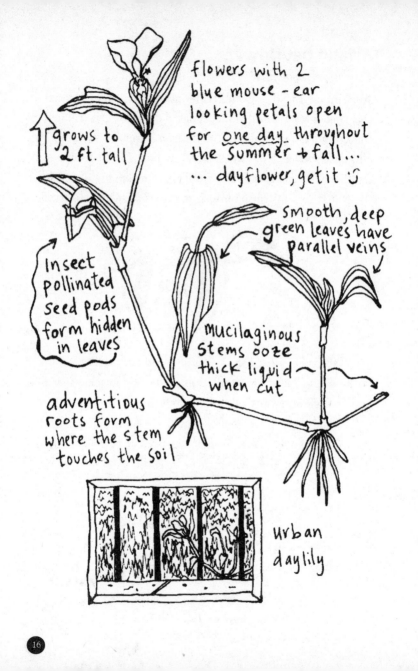

The bright blue petals of the Asiatic dayflower beckon us to look closer, as this color is not found so often in nature. Bees come in close, too, because the bright flower signals nectar. This dayflower that shares its beauty with us for just one day, as the name suggests, is a welcome sight in the gray of the city.

Humans have long been drawn to this plant. In its native China, tea is made from the leaves to soothe sore throats or as a throat gargle. Its leaves are edible and can be eaten raw in a salad or cooked in soups or stews. The blue petals are the source for a bright blue dye. In fact, Asiatic dayflower provided the initial blue dye used to color the famous Japanese ukiyo-e style woodblock prints before a more permanent blue was imported from the West!

Asiatic dayflower found a home here in the Northeast and South and is slowly spreading across the US, though there are currently no reports of sightings in some Western states. It spreads and outcompetes other plants by rooting at the stems where they touch the ground, which helps to form huge colonies.

Seeds are eaten by songbirds. There is evidence that this plant can be useful in phytoremediation efforts, especially its ability to remove copper from the soil. Asiatic dayflower feeds and cleans our city, and it does so while looking damn good too.

Black Nightshade

Solanum nigrum

Black nightshade is an annual of the Solanaceae, or Nightshade, family. It is related to many well-known and much-used plants: tomato, potato, eggplant, petunia, cayenne pepper, and tobacco among them.

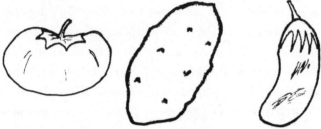

Habitat: Black nightshade prefers sunny and moist soil but does not discriminate. It will grow in many types of disturbed places. You will find it in sidewalk cracks, vacant lots, public parks, rubble heaps, and landscaped areas.

Black nightshade traveled from Eurasia, and it is unknown how it was introduced. Maybe by way of a cow's hoof? It is unknown exactly when black nightshade reached North America, but folks collected some on the West coast of Washington state as early as 1825.

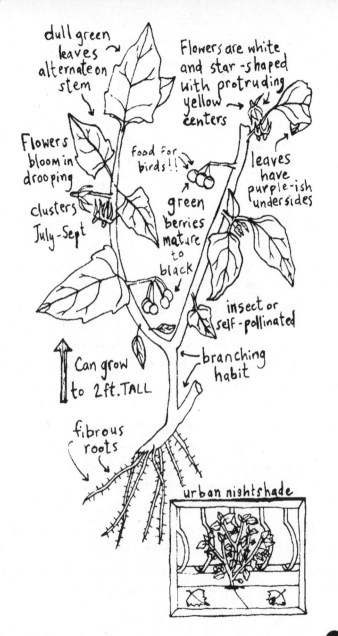

Black nightshade gets its name from the color of its small juicy berries and from the family it belongs to. Its genus name, *Solanum*, is derived from the Latin word solamen, which means "quieting." This refers to the narcotic effects of some of the plants in this genus. I'm sure it took the lives of a few unfortunate ancestors for us to learn that the quieting effect is permanent in some cases. Deadly nightshade, for instance, is a similar-looking cousin with purple flowers whose quieting effect is severe, which provides me the opportunity to remind you, dear reader, to please be 100 percent sure before ingesting any foraged plant.

Black nightshade leaves, stems, and immature green berries do contain the compound solanine, which is toxic to humans, so depending on where you live in the world, the leaves are enjoyed boiled like spinach or, conversely, children are taught early on to avoid it. Once the berries mature and turn black, they are no longer toxic as the proportion of solanine is less, and these have been safely turned into pies and preserves.

Many species of the Solanaceae family have been used for their hallucinogenic, medicinal, and poisonous properties. Black nightshade is no different, having been used in traditional medicine in Europe, India, Asia, and Africa to treat inflammations, skin disorders, and eye problems.

Black nightshade is visited by many different species of bees and birds, and we get to enjoy the funky shooting star–shaped flowers. Seeds remain viable in the soil for many years and can sprout under a variety of conditions, solidifying this plant's spot in the urban world.

Bladder Campion

Silene vulgaris

A short-lived perennial of the Caryophyllaceae family, aka the Pink or Carnation family, bladder campion is related to the popular garden flowers mountain sandwort, sweet William, baby's-breath, and dianthus.

Habitat: Prefers sandy or gravelly soils in full sun. Can be found in vacant lots, rubble dumps, railroad rights-of-way, landscaped areas, and public parks.

Came from Europe, Asia, and North Africa. Probably introduced as an ornamental plant because of its flowers in the late eighteenth or early nineteenth century. Escaped cultivation and naturalized across North America.

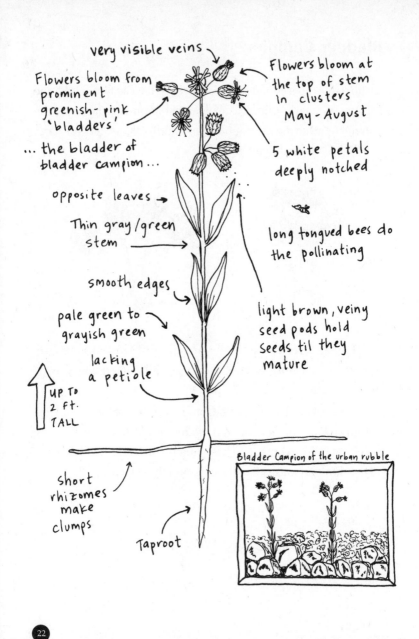

Vulgaris, the species name, means "common" in Latin. Bladder campion was a constant sight throughout its native lands. And though it is not native to this land, it has become common here too.

The flowers are really striking; I can understand why early Europeans wanted to plant it in their gardens. When it graces my garden, I let it stay. I even transplanted a bladder campion plant once. It is thriving. The seeds are wild looking, like funky nautilus shells, something from the sea. And the seed pods, too, have character, resembling faded ancient vessels. Bladder campion offers something to look at during every stage of development. Like other members of the pink family, the foliage contains some saponins, not enough for it to be used for soap but just enough to make herbivores avoid eating it, though livestock have been known to eat the foliage in overgrazed pastures. It is known colloquially in Spain as collejas, and the young shoots and leaves are enjoyed as a cooked green.

White campion is a look-a-like that grows in similar conditions, but its bladders are not as prominent and it has hairy, deep green stems and leaves. The flowers of bladder campion are so interesting you can tell that Mother Nature had fun while making them. They are a living example of the importance of injecting life with fun and playfulness even in the hard-to-root- down sandy soil.

Broadleaf Plantain

Plantago major

An herbaceous perennial of the Plantaginaceae family, the Plantain family. *Plantago* is the largest genus of this family and different species of *Plantago* are found spread throughout the temperate regions of the world.

Habitat: Grows best in moist, nutrient-rich soil. Can tolerate salty and dry conditions as well as mowing and trampling. Found in vacant lots, tree pits, urban meadows, gravel lots, and pathways, roadsides, median strips, public parks, lawns, landscaped areas, and building foundations.

Broadleaf plantain is from Europe. It's unknown how it arrived in North America, but it came here early on with Europeans. Broadleaf plantain's presence was recorded in New England in 1672 and Canada in 1821.

Said to be called "white man's footprint" by many Native folks as it grew wherever European colonizers cleared the land.

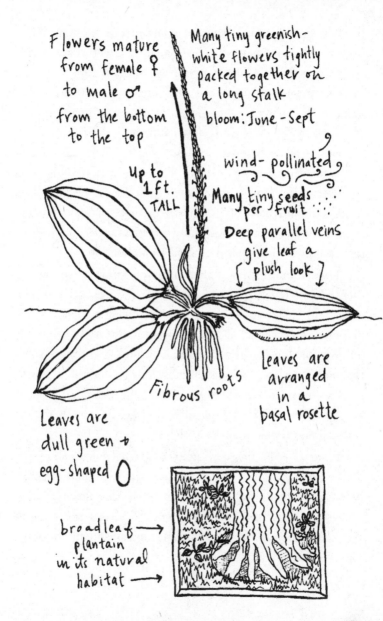

The genus name *Plantago* comes from the Latin for "sole of the foot," named so because of the leaf shape. How interesting that the name given by Native peoples means something very similar. It is amazing, too, how this herb was used for many of the same ailments by cultures around the world, which science is now confirming with laboratory studies. Broadleaf plantain does indeed have medicinal properties that can support our skin, respiration, digestion, and overall immunity. It was one of the sacred herbs in the Anglo-Saxon Nine Herbs Charm because of its amazing healing qualities and is best known for its use as an antidote to pain and itching caused by rashes and bug bites.

This is another plant that once you know it, you see it everywhere! I just got back from a neighborhood walk—I needed to revive my writing brain—during which I spied plantain on every block. Some were large and lush with pillowy looking leaves you could sink into, and others were tiny and haggard looking, humbly displaying the trait of morphological plasticity: an adaptation in which plant growth reflects the soil vibrancy—small and scraggly growth in poor soil, tall and robust growth in rich soil.

It has a hard time growing in compacted soil, as all plants do. Soil compaction collapses air and water tunnels, making it tough for plants to drink and breathe. As a service to the ecosystem, broadleaf plantain's tough fibrous roots break through compacted soil, making it more hospitable to other plants.

A plant so helpful and giving to other beings, perhaps it is its commonness that prevents us from truly seeing what is there. I think it's time we rejoice that this medicine can be found in so many places, instead of lamenting its spread.

Burdock

Arctium minus

Burdock is a biennial member of the Asteraceae family. Related to yarrow, daisy, galinsoga, calendula, artichoke and MANY others.

Habitat: Loves moist, rich soil and full sun or part sun. Can be found in public parks, vacant lots, foundations of buildings, rubble piles, along waterways, median strips, and railroad rights-of-way.

Burdock came from Eurasia. It is unclear how or when it was introduced, though it was introduced sometime in the 1600s and reported as a common New England weed by the 1700s. Perhaps it came by way of an animal . . .

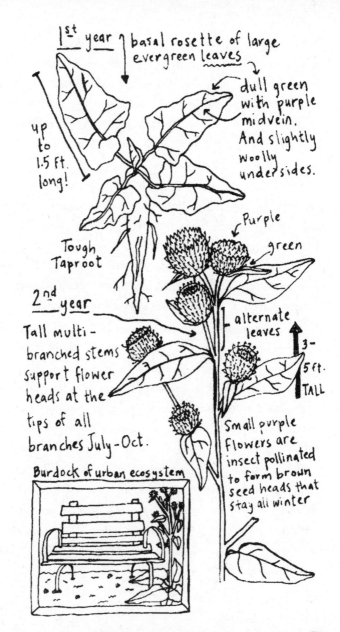

Seed heads are covered in tiny hooks, aka burs, an adaptation developed to help burdock spread using animal fur as a vehicle. This amazing adaptation caught the eye of George de Mestral, a Swiss engineer, who, by capitalizing on the concept of biomimicry (while hiking he noticed the burs stuck to his clothes as well as his dog's fur and investigated how), invented Velcro in 1955!

In its native lands of Europe and Asia, burdock has long been used as a medicinal herb. The root contains a compound called inulin, which feeds our beneficial gut flora. It is rich in lots of vitamins and minerals and supports the digestive system, endocrine system, and overall immunity. The root is dried and taken in tea form. The leaves have a cooling effect and are placed on the body to relieve skin ailments.

Common burdock spreads by seeds (far and wide, thanks to its burs). Pollination is ensured by the many insects that drink its nectar. Long-tongued bees such as bumblebees, honey bees, miner bees and leaf-cutting bees, as well as some butterflies are frequent guests. The great number of beings that come to feast reveals that common burdock supports a robust ecosystem.

Burdock thrives in urban environments. The long and strong taproot can dig deep into the soil to draw out water and nutrients. What common burdock doesn't use stays close to the surface and is made more available to other plants with shorter root systems—acknowledging the web of life by providing for others.

Canada Goldenrod

Solidago canadensis

A perennial of the Asteraceae family, or the Sunflower family. Related to many familiar plants including: dandelion, daisy, New York ironweed, yarrow, coneflower, and many more.

Habitat: Grows best in well-drained soil in full sun. Can be found in vacant lots, building foundations, rubble dumps, drainage ditches, railroad rights-of-way, and public parks.

Comes from Eastern North America.

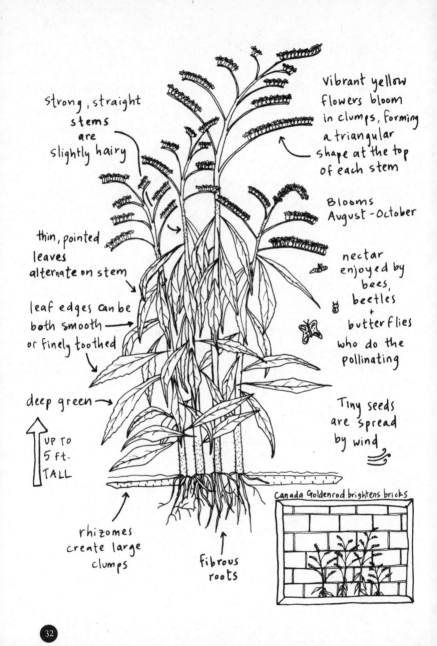

Canada goldenrod is an unassuming tall mass of foliage before the flowers arrive, but when they do, wow, we are lucky to witness. The flowers are blooming when the first trees are starting to change into their autumn colors so they add to the beautiful symphony of yellows, oranges, reds, and purples blaring on the city streets. Canada goldenrod is considered a welcome addition to the city by the loads of insects, including bees, beetles, and butterflies that feast on the nectar. It has been able to carve a home here in the urban ecosystem thanks to its tough roots and spreading rhizomes as well as its tolerance to the ever present urban pollutants.

Before the rise of the modern city, folks found it growing in fields and grasslands with other wildflowers. Many pre-colonial Native traditions use Canada goldenrod to ease a variety of conditions: fever, pain, and sore throat. Beyond these uses, the Iroquois, in particular, find it helpful for such disparate uses as helping babies sleep and increasing gambling luck. A pretty good plant to have around, I'd say. Canada goldenrod was introduced to Europe in 1645 and China in 1930 and is now considered an invasive species in both Europe and Asia.

Canada goldenrod is often wrongfully accused of causing hayfever, but it is actually the less conspicuous flowers of ragweed that cause the sneezing, itchy eyes, runny nose, and post nasal drip in the allergy sufferers. What else is Canada goldenrod wrongfully accused of? It certainly should not be accused of invading our cities. We built our cities atop its home, and Canada goldenrod has found a way to live with the changes. We could stand to let it find a way into our hearts.

Chickweed

Stellaria media

An annual of the Caryophyllaceae or Pink family, a large family that includes carnations.

Habitat: Loves sunny, nutrient-rich soil but takes advantage of other growing conditions as well. Can be found in vacant lots, public parks, unmowed lawns, rubble dumps, sidewalk cracks, building foundations, and landscaped areas.

Came from Eurasia. A widespread plant that can be found from high altitudes in the tropics to high latitudes in the north. It's unknown how it traveled: perhaps snuck in a bag of seeds?

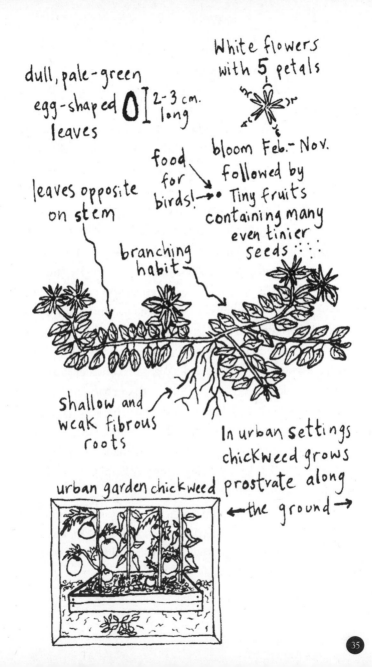

Chickweed is so cute! Its flowers look like little twinkling stars, which is why the genus is called *Stellaria*, the Latin word for "star." Five pearly-white, notched petals call to us to look closer, and when we do, we are treated to a circle of pollen-laden stamens shining bright yellow. It is definitely worth bending down for. Thankfully, there's a long flowering season to peep the flowers. Chickweed is a winter weed so prefers to grow in the cooler temperatures of spring and fall and, if it feels right, chickweed seeds can germinate in the fall and then take advantage of any mild days in winter to grow!

In optimal conditions, chickweed can mature and set seed in just over a month, which is how it is capable of having up to five generations in one year. Each generation can form large mats on the ground because of the roots produced at the stem nodes when they touch the ground. When the conditions are right for the seeds to germinate, our cities are blessed with a great amount of chickweed. We can consider ourselves lucky that chickweed is so prolific in the urban environment because it is such a delicious and nourishing tonic. It is a powerful tissue healer that can deliver its nutrients as a salad green or as a juice for internal relief or made into a poultice for external relief. In Victorian times in England, chickweed was added to fancy sandwiches. So fancy!

The bright green, juicy foliage looks yummy, and it is. Chickens love it too, hence the common name chickweed. A mat of chickweed on the ground is a joyful reminder of the abundance that surrounds us.

Chicory

Cichorium intybus

A perennial of the very large Asteraceae family, the Sunflower family. Other examples include mugwort, coneflower, yarrow, and New York ironweed.

Habitat: Does well in sunny, dry places and can stand an elevated pH (basic, limestone-y soil). Can be found in vacant lots, gravel paths, roadsides, rubble dumps, building foundations, landscaped areas, median strips, and sidewalk cracks.

Came from Eurasia. Maybe as a stowaway in a bag of grains?

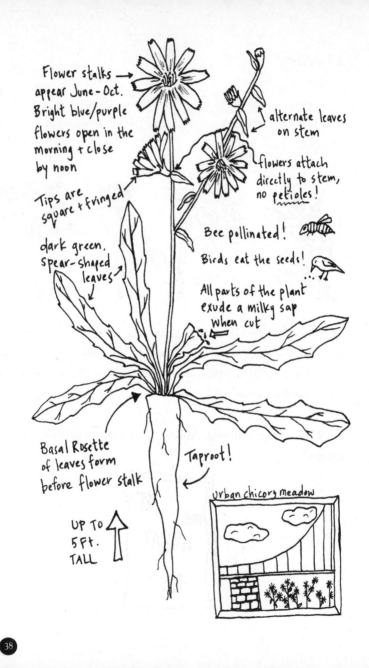

Chicory has lived in cities alongside humans for a long time. In fact, there is record of it being cultivated in Ancient Rome and Egypt! And for good reason—not only is it low maintenance, but the leaves, when blanched to remove bitterness, are a delicious and nutritious salad green or potherb. The legacy of chicory's revered status in Ancient Egypt lives on in its name, which may have come from a corruption of the Arabic word for the plant, *chicourey*.

Besides being enjoyed at the dinner table, the leaves and roots of chicory were used as traditional medicine throughout Europe to treat liver problems and gout. In Egypt, chicory was cultivated as an herb to purify the blood and liver. When chicory arrived in North America, the Cherokee people made an infusion of the root as a tonic for the nerves. Nowadays, a fun way to drink the root is to roast, ground, and brew it as everybody's favorite decaffeinated coffee substitute.

Chicory is a great asset to the urban ecosystem. Short-tongued bees and other insects enjoy chicory's offerings. And there is evidence that chicory, like others in the Asteraceae family, can be used in phytoremediation efforts to clean contaminated land.

The lovely, pale lavender flowers of chicory can transfix even the most busy body, allowing us to slow down, forget the clock, and flow with the wind.

Cleavers

Galium aparine

An annual of the Rubiaceae family, the Madder family. A relative of the plant that gives us coffee as well as the dye plant called madder and the ornamentals gardenia and sweet woodruff. Other fun common names include: catchweed bedstraw, stickywilly, cleaverwort, and robin-run-the-hedge.

Habitat: Thrives in many types of soil and shade or sun. Can be found in landscaped areas, public parks, pathways, and along fences.

It is unclear where it first came from, perhaps Eurasia, but it is now naturalized in North America, Europe, Asia, and Australia. Perhaps brought unwittingly by a house cat.

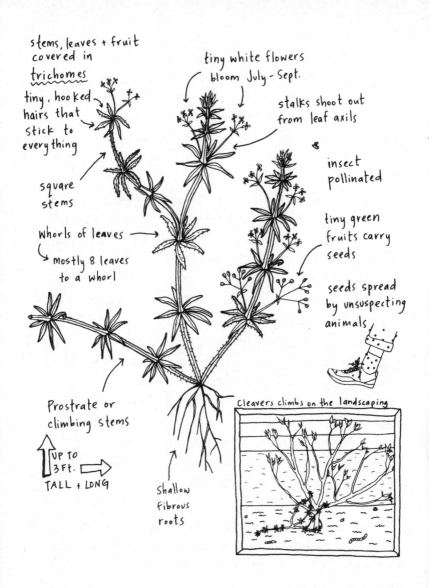

The species name, *aparine*, comes from the Greek word meaning "to seize," and that is what it does best. Cleavers will seize any opportunity to climb on a plant in its path to get closer to the sun. It will make use of every passing animal, its seed pods seizing upon bodies as transport for the next generation. It is this opportunistic behavior that has carried cleavers to all the parts of the world with the suitable temperate climate it loves.

If you find yourself covered in seeds as the chosen animal, don't get mad, make coffee! Well, a lightly caffeinated coffee substitute; it is a close relative after all. Historically, cleavers was eaten raw as one of the first spring greens, and it was considered a spring tonic helping to invigorate the body after winter. Herbalists of various traditions make a tincture for stimulating lymphatic drainage and consider it useful for the skin and urinary system, including the prostate.

Geese and chickens love to eat the stems and leaves, which is how it earned another common name, goosegrass. Insects love to visit the flowers for nectar, and I don't know what they call cleavers, but word gets out when cleavers is in bloom because the plants are abuzz with activity.

The stickiness of the plant provides the young and the young-at-heart much opportunity for play. I have bedazzled many shirts with a design of seed pods.

Cleavers offers so much to the urban ecosystem, and its leaf arrangement adds texture to the landscape. Let's acknowledge its value and stick up for cleavers by saying its name. It's only fair since cleavers has stuck by our side for so many years.

Coltsfoot

Tussilago farfara

A perennial of the Asteraceae family, the Sunflower family. Related to ornamental knapweeds, succulent senecios, and weedy New England hawkweed.

Habitat: Likes moist areas in full sun or dappled shade. Can be found in vacant lots, drainage ditches, public parks, median strips and along waterways.

Came from Europe. Likely introduced early on by European colonizers for medicinal purposes.

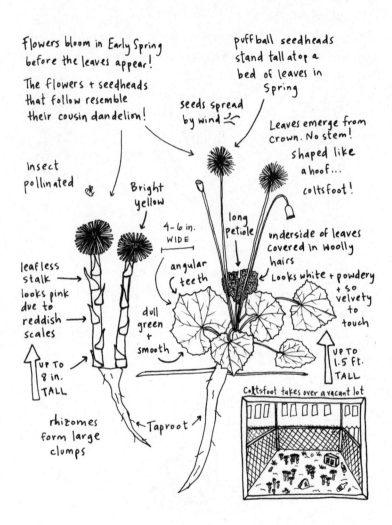

It took me a long time to truly get to know coltsfoot. I could recognize the leaves and seed head on walks, but I never saw the flower. From the clump of large, hoof-shaped leaves, a long stalk topped by a puffy seed head emerges in late spring, but the flower eluded me. I had no idea that what I saw on early spring walks—these clumps of solitary, pinkish stalks emerging straight from the soil, no leaves to be seen and topped by dandelion-looking flowers—were the flowers of coltsfoot. They looked almost like the offspring of dandelion and ghost pipe, were they able to cross-pollinate. When I finally learned (later than I'd like to admit) that those were the flowers of coltsfoot, my mind was blown. For one thing, I was so happy to have an answer to this riddle. For another, I was so impressed by the ingenuity of this plant.

Over the winter, the rhizomes of coltsfoot store food as carbohydrates, which supports the growth of the early spring flowers that bloom before the leaves emerge. The bees and flies that drink the nectar are so grateful for such an early blooming flower, fresh food for their hungry tummies. Historically, humans were always happy to see the leaves. They were used to treat a wide range of chest complaints in their native lands. And practicing herbalists today call it an amazing respiratory ally and brew the leaves in tea or dry them for a healing smoke.

We don't see too many herbaceous plants that flower before the leaves come out, making coltsfoot a fun addition to the urban ecosystem. Coltsfoot, along with the food and medicine it brings, teaches us that there are many ways of doing things.

Common Blue Violet

Viola sororia

A perennial of the Violaceae family, the violet family. At least half of the nine hundred species in the Violaceae family belong to the genus *Viola*–that's a lot of violets!

Habitat: Grows best in cool, moist, shady soil but can tolerate full sun, dry soil, and compaction. Can be found in public parks, landscaped areas, tree pits, sidewalk cracks, lawns, and drainage ditches.

Comes from Eastern North America.

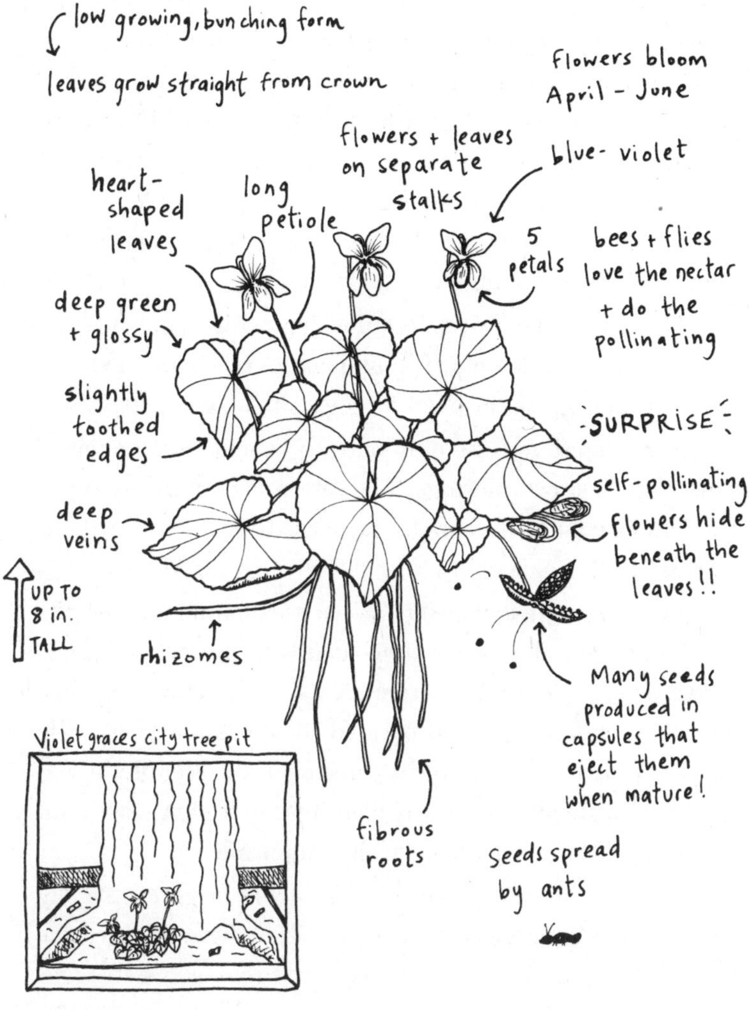

- low growing, bunching form
- leaves grow straight from crown

Flowers bloom April – June

flowers + leaves on separate stalks

blue-violet

heart-shaped leaves

long petiole

5 petals

bees + flies love the nectar + do the pollinating

deep green + glossy

slightly toothed edges

-SURPRISE-

self-pollinating flowers hide beneath the leaves!!

deep veins

UP TO 8 in. TALL

rhizomes

Many seeds produced in capsules that eject them when mature!

Violet graces city tree pit

fibrous roots

Seeds spread by ants

Common blue violet turns the monochrome green lawn into a canvas for its stunning display of spring flowers that range in color from deep violet to velvety white. The fun flowers are edible and can bring the same pop of color to a salad or to a mixed drink when frozen in ice cubes. Common blue violet is not just a looker but a healer as well. The leaves are cooling and healing to our bodily tissues. They can be eaten raw in a salad or added to a green smoothie. A real nourishing herb, common blue violet is very high in protein, iron, calcium, and Vitamins A and C.

Thanks to their deep relationship with this land, the Cherokee learned the nourishing qualities of common violet, and continue to use the plant for a variety of ailments.

Humans are not the only ones to benefit from common blue violet's sweet offerings. The caterpillars of many types of Fritillary butterflies are frequent visitors of violets, there to feast on the foliage.

An interesting and lesser-known use of common violet is by the tiny ant. The seeds are coated with protein and lipid-rich morsels, a delicious ant treat. Ants enthusiastically gather the seeds and bring them back to their nests to feed their larvae. When the seed coat is consumed, the seed is thrown onto the waste pile, effectively planting the seed. A wonderful example of a mutually beneficial relationship.

Back to the spring flowers, they truly brighten the day and any street on which they bloom. They have survived the onslaught of tar and cement, and for it they grow stronger and brighter. They know what it is to come back from the rubble. Sit with them and let them help to soothe any heartbreak.

Crown Vetch

Coronilla varia

A perennial of the Fabaceae family, the Pea family. Related to red clover, honey locust, white clover, and all of our edible bean plants.

Habitat: Loves sun and is tolerant of salt, compacted soil, and low-nutrient soil. Can be found in landscaped areas, roadsides, median strips, highway embankments, gravel paths, and public parks.

Introduced to the US from China by the USDA to serve as erosion control on highway embankments and for revegetation of former mining sites. Used extensively in the 1950s and quickly escaped cultivation.

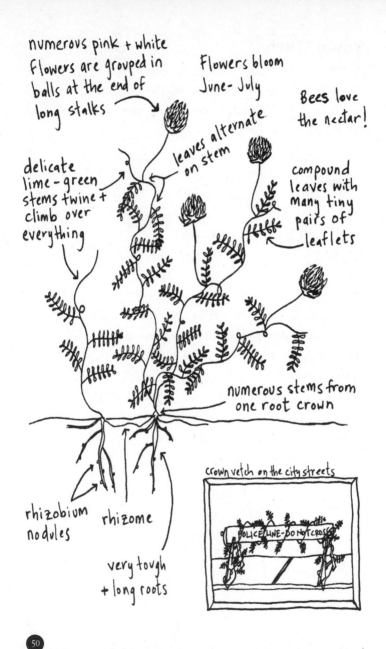

It is no wonder that crown vetch escaped cultivation; its rhizomes can be up to ten feet long and an individual plant can spread seventy to one hundred square feet in four years. Bees love the nectar and ensure that crown vetch flowers are pollinated to form seeds of the next generation.

Crown vetch is well loved by many animals of the urban ecosystem and beyond. Many insects feed on the foliage, including the larval phase (caterpillars) of several species of butterfly. Deer, elk, and livestock enjoy the foliage, and the dense mats on the ground create protective cover for ground nesting birds, meadow voles, and rabbits. Crown vetch has not become a salad staple for us because all parts of the plant are poisonous to horses and humans due to the presence of nitroglycosides.

Like other members of the Fabaceae family, crown vetch works with rhizobium bacteria to fix nitrogen from the air into the soil as ammonia. This process essentially fertilizes the soil and feeds the other plants that share that land.

Crown vetch feeds, fertilizes, and serves as a safe space, its presence sustaining a diversity of life within the urban ecosystem.

Curly Dock

Rumex crispus

Curly dock is a perennial of the Polygonaceae family, the Buckwheat family. Related to well-loved sorrel, buckwheat, and rhubarb, as well as lesser-loved Pennsylvania smartweed and Japanese knotweed.

Habitat: Prefers moist, clay soils in full sun. Found in vacant lots, rubble dumps, construction sites, landscaped areas, fields, and bioswales, as well as along railroad tracks, pathways, and median strips.

Curly dock came to North America from Europe. It is now found throughout the US.

The plants of the genus Rumex can be found in many, many places, and *Rumex crispus* is the most prevalent of them all, found the world over (minus Antarctica).

It is unknown whether it was brought on purpose or not. Perhaps it came by boat?

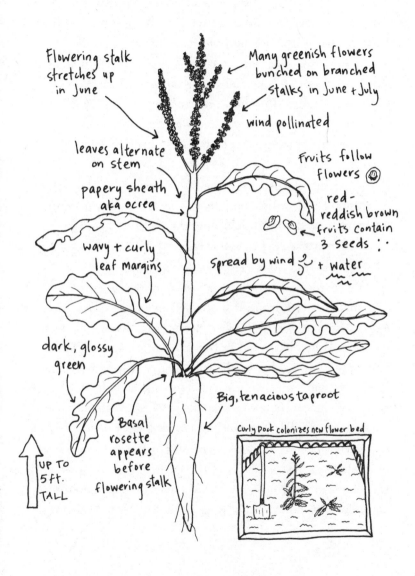

The species name *crispus* is Latin for "closely curled." A nod to a key identifying feature, its curly leaves, which distinguishes it from its multitude of cousins.

In the urban environment, curly dock fills a niche that few other plants can. Its long and strong taproot, which grows up to three feet, digs deep into even the most compacted soil. This creates pathways for air and water and makes the land more hospitable for other plants. Curly dock is definitely a team player.

Many city birds enjoy the seeds, including the song sparrow and red-winged blackbird. Insects chow down on the leaves, stem, and flowers, including the larvae of many moths and the American copper butterfly.

Humans too have long benefited from curly dock, specifically the root, which has a history of use as a medicinal herb in its native Europe. It was called on for use as a laxative and astringent, as well as to treat lung, liver, and skin ailments. When it arrived in North America, several Native tribes incorporated it into their healing regime as well. The leaves are high in vitamins and minerals, and they can be eaten, though do be aware that the leaves are also high in oxalic acid, so eating too much can cause kidney damage.

Curly dock seeds can remain viable for sixty years or more. If even just a part of the root is left in the ground, the plant can resprout. Tolerant of compaction and drought, curly dock has found many ways to carve a home out in the city.

Somehow, a stalk of curly dock in winter—standing straight and tall with its reddish brown seed heads illuminated against the gray landscape—has a way of both representing the past and future, all while I remain firmly in the present struck by its solitary beauty.

Daisy Fleabane

Erigeron annus

Another annual of the Asteraceae family, the sunflower family. Related to chamomile, calendula, and sunflower.

Habitat: Loves fertile soil in full sun but is also very drought tolerant. Can be found in public parks, median strips, sidewalk cracks, building foundations, rubble dumps, gravel paths, and railroad tracks.

Comes from Eastern North America.

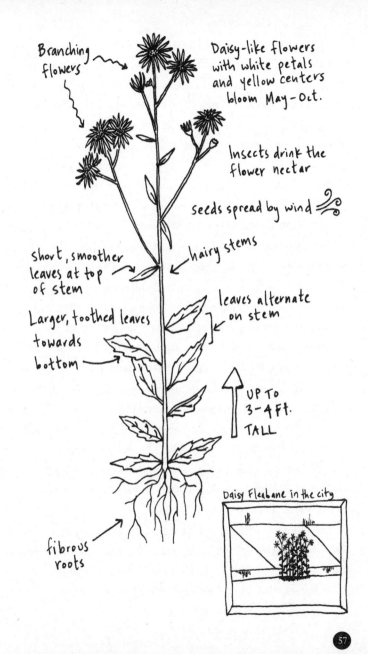

The branching flowering stalks of daisy fleabane means we are graced with a profusion of small, daisy-like flowers throughout the growing season here in Northeastern America. The flowers have a shaggy look, and the stems and leaves carry a little peach fuzz, similar to an old man's whiskers, which are probably what the taxonomists were thinking of when they named this plant. The genus name *Erigeron*, comes from two Greek words, one meaning "early" and one meaning "old man." "Early" refers to the early blooming flowers, and "old man" to the shared trait, whiskers.

The flowers attract and feed a variety of insects, as do the leaves. Lean in close and you can see them feasting. You won't catch gnats or fleas hanging around, however. The common name, fleabane, comes from the belief in its ability to rid an area of gnats and fleas when burned. There are many species of the genus *Erigeron* that are native to this land and each has a long history of use by many Native tribes.

Thanks to its prolific flying seeds, daisy fleabane is now considered a weed in Europe, reversing the usual direction of invasive plant travel. It has also slowly been advancing toward the Western coast of the US, enlarging its American range. Despite its weedy status, cultivated varieties are sold in nurseries and are promoted as easy-to-care-for, low maintenance plants.

The daisy fleabanes that we find popping up around us in our urban home sure are easy to care for and low maintenance. And even better, they're free! Saying daisy fleabane allows us to better see the daisy-like beauty that surrounds us.

Dandelion

Taraxacum officinale

A perennial of the Asteraceae family, the Sunflower family. Related to marigold, artichoke, chrysanthemum, and lettuce, among many many others.

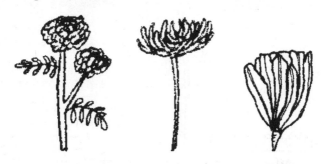

Habitat: Loves full sun and nutrient-rich soil. Can tolerate some shade, trampling, compaction and drought. Can be found in lawns, vacant lots, landscaped areas, sidewalk cracks, median strips, building foundations, and so many other places.

Came from Eurasia. It was introduced by Europeans in the seventeenth century for use as a potherb and medicine.

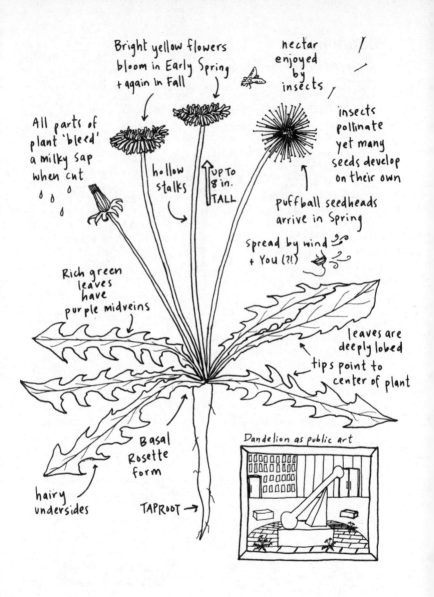

Dandelion is so common that it's easy to miss here in North America. Its edibility and healing qualities were not missed in its native Europe and Asia, however. In traditional Chinese medicine dandelion is defined as cold in nature and used to benefit the liver and stomach. The species name, *officinale*, is Latin for "medicinal or culinary use," so it was also clearly being enjoyed by Europeans. Even in its early days in North America, dandelion was prized for its health benefits. Twenty-six Native tribes are documented to have used dandelion as food and medicine, from East to West, North to South. As the dandelion traveled across the continent, folks were quick to learn its attributes and healing power.

Somewhere along the way, we lost sight of the dandelion. Perhaps it was the invention of the lawn? Americans looking for the "perfect lawn" dump nearly 90 million pounds of herbicides and pesticides on lawns each year. The sweet natured, joyful dandelion is cited as the cause of the spread of the use of herbicides on lawns. Really, folks should feel lucky if they have a profusion of dandelions growing on their lawn, because it's a signifier of fertile soil. It is higher in Vitamin A than any other plant, and it is a feast for the bees because of the many flowers packed together. This plant offers sustenance and more.

My young neighbor likes to rub the flowers on her arms and paint her skin yellow. My two-year-old-son can recognize dandelion, and it is the only flower name he knows how to say. Sure, you have to use your imagination a little to hear "dandelion" when he says it, but he is so happy to know and recognize this plant that's all around him. When we are taught to say the name of dandelion and allow it to bring us joy, our world is immediately filled with love.

Evening Primrose
Oenothera biennis

A biennial of the Onagraceae family, the Evening Primrose family (one of the smaller plant families). This plant's best known cousins are ornamental plants gaura and fuschia.

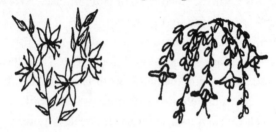

Habitat: Grows best in dry, sandy, gravelly soil in full sun. Can be found along fences, building foundations, vacant lots, sidewalk cracks, rubble dumps, landscaped areas, median strips, and public parks.

Comes from Eastern North America.

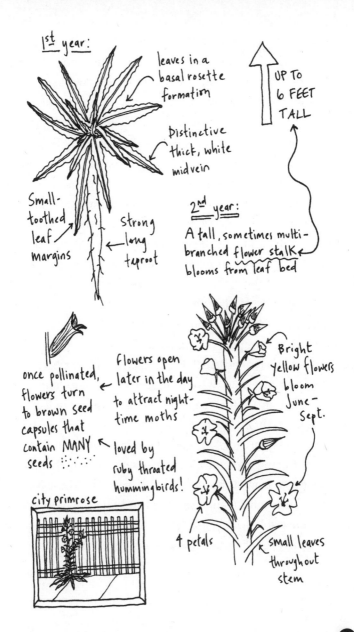

The botanical name of evening primrose gives us a clue to its life cycle: *biennis* comes from *biennial*, meaning "lasting two years." The first year, it is easily overlooked as a basal rosette of leaves blanketing the land. The second year, however, when the flower stalk rises? What a showstopper!

I can just picture the soft yellow flowers mingling with other wildflowers, sticking out amongst the grasses of fields and prairies in its native habitat. Evening primrose is native to this land, but it was not born of the city. It moved here and hustles just like the rest of us. In the city, evening primrose supports a diversity of life, including night-flying moths, bees, and ruby throated hummingbirds, beings whom it has been in relationship with for centuries, feeding them nectar and a small taste of what once was.

Evening primrose continues to move and now has naturalized in Europe. Folks have made good use of evening primrose. Modern herbalists use it as a demulcent herb that moistens our nervous system, digestive system, and skin, as well as an anti-inflammatory that can ease the effects of rheumatoid arthritis. Before traveling to Europe, evening primrose was already in deep relationship with Native Americans, being used for a variety of ailments, including as a stimulant to combat laziness by the Iroquois! We need not fear that this beautiful and healing plant will disappear. Seeds can remain viable in the soil for up to seventy years as they wait patiently for the next disturbance.

Galinsoga

Galinsoga quadriradiata

An annual of the Asteraceae family, the Sunflower family. Another relative of mugwort, liatris, zinnia, and lots of others.

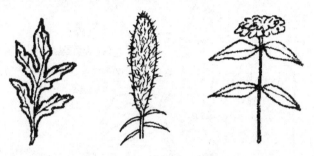

Habitat: Seeds sprout in sunny, disturbed sites. Tolerant of compaction and drought.

Came from Central America. Migrated on its own!

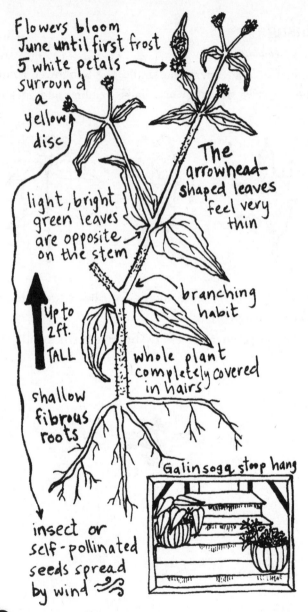

Flowers bloom June until first frost 5 white petals surround a yellow disc

The arrowhead-shaped leaves feel very thin

light, bright green leaves are opposite on the stem

branching habit

Up to 2 ft. TALL

whole plant completely covered in hairs

shallow fibrous roots

Galinsoga stoop hang

insect or self-pollinated seeds spread by wind

Despite sometimes going by the name "gallant soldier," galinsoga is so diminutive and fragile. It looks easy to pull out of the garden bed (which it is) and nothing like a gallant soldier, so how did this other common name come about? Galinsoga was named for Spanish botanist Don Mariano Martinez de Galinsoga, a director of the Botanic Gardens of Madrid in the eighteenth century. When the plant galinsoga landed in London, residents had a hard time pronouncing its given name and changed it to gallant soldier.

Galinsoga moves fast! It can go from sprouting to seed production in three to four weeks. This is how it is able to pack up to three generations in one growing season. This quick movement has enabled it to spread around the world. It can now be found across North America, Europe, Asia, and North Africa. A few years ago, I saw it growing in Northern India and bubbled over with excitement, so happy to see a familiar face. The young resident I was walking with did not share my enthusiasm. He told me with disdain of its weedy status and how it was "good for nothing."

Colombianos would disagree with him; they call this plant *guascas*, dry the leaves and stems, and use it as a popular seasoning. It is also enjoyed as a green vegetable in soups and stews in its native land of Central and South America.

So here we have an issue of perception. Sometimes all it takes is for us to consider a plant valuable in order to remove the hated label of weed. But what if being alive is valuable enough?

Garlic Mustard

Allaria petiolata

A biennial of the Brassicaceae family, the Mustard family. Related to many familiar plants such as nasturtium, horseradish, cabbage, etc.

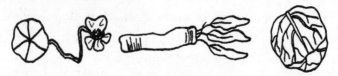

Habitat: Thrives in woodland understories in dappled shade but can be found growing in many other conditions: roadside clearings, landscaped areas, and public parks.

Came from Eurasia. Unknown whether on purpose or not, but most likely introduced as a potherb in the early 1800s.

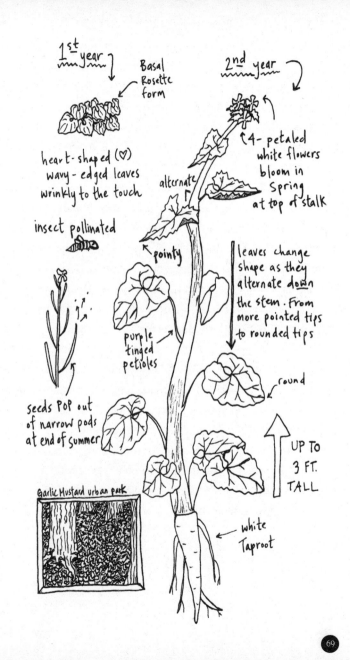

Garlic mustard, as the name suggests, has a strong garlic flavor that has been enjoyed in Europe as a spring green for years. It is quite nutritious, high in Vitamins A and C, and the leaves are believed to strengthen the digestive system. We Americans tend to turn our noses up at the strong garlic scent (as do deer, by the way), so we have yet to benefit from the nutrients garlic mustard offers.

Garlic mustard is considered to be a tough competitor, and conservationists let out a collective groan when it is found growing in a forest understory. One trait that gives it an edge is that it emits chemicals that inhibit the growth of beneficial mycorrhizal fungi growing on the roots of other plants, an action that keeps those plants small and gives more room for garlic mustard to move. If you try to manage its spread by cutting off the flowering stalks when the seeds are immature, the seeds are able to mature on the cut stalk— pretty amazing! This is not a plant to add to the compost pile.

Instead, try making homemade paper from the stalks and leaves of garlic mustard or simmer the whole plant in water to produce a yellow dye. Garlic mustard can feed our body and our creative soul. Perhaps it is tired of being overlooked and its prolific spread is a call for us to be in right relationship. If we listen and take advantage of the abundance in front of us, we can help to make room for more plant beings.

Green Foxtail

Setaria viridis

An annual of the Poaceae family, the Grass family. Cousin to crabgrass and many yummy staples such as oats, wheat, rice, corn, barley, and sugarcane.

Habitat: Tolerant of salt and compacted soil and loves full sun. Can be found in sidewalk cracks, building foundations, gravel paths, public parks, median strips, and landscaped areas.

Came from Eurasia. Brought intentionally as forage grass for livestock in the 1800s.

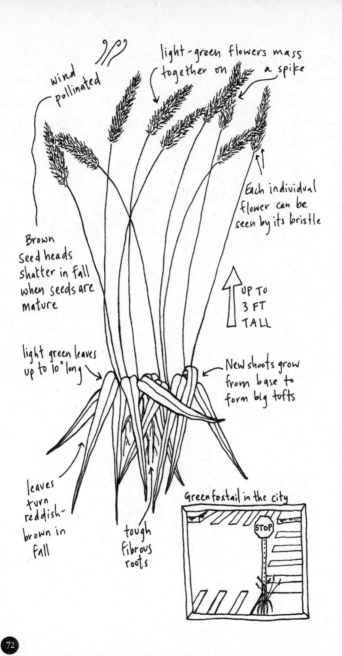

Green foxtail followed the cows to North America. It has since migrated from grazing land to our urban centers because of its tolerance to the challenging conditions of city life. Green foxtail traveled to the big city quickly thanks to its fast maturation rate, a trait that enables it to have multiple generations in one growing season.

Living alongside humans in the city is nothing new for green foxtail, as it has been living with humans for a very long time. Green foxtail was domesticated in China about 8,000 years ago. The result, foxtail millet, remains the most grown millet in Asia. Green foxtail seeds are edible. Though the seeds are scattered and dispersed upon maturity, the domesticated variety of this plant, foxtail millet, was cultivated to retain its seeds upon maturity, making it easier for harvesting. Green foxtail seeds are still turned to as a food source in starving times throughout the world. And when there is no chicory around, you can roast green foxtail's seeds as a coffee substitute. There is documentation of a long history of medicinal use as well, mainly for treatment of fever and bruises and to induce urination.

Green foxtail plays its part in the urban ecosystem: many insects feed on the foliage and birds enjoy the seeds. They and we are lucky that green foxtail is able to withstand city life. Green foxtail is a tough plant whose green, bristle-like flowers add a softness to the hard-edged urban landscape. These plants invite us to open our eyes and see that there is strength in softness.

Ground Ivy
Glechoma hederacea

A perennial of the Lamiaceae family, the Mint family. Many famous cousins, including basil, peppermint, oregano, rosemary, thyme, lemon balm, skullcap—half the herb garden!

Habitat: Grows best in shady, moist conditions. Tolerates full sun, as well as some drought and trampling. Can be found in public parks, lawns, landscaped areas, and highway drainage areas and alongside waterways.

Came from Eurasia. Believed to have been brought to North America early on for medicinal purposes.

Ground ivy is quick to fill in the bare spots of soil where our cultivated plants have perished and loves to quietly (and quickly!) creep across lawns. If we choose to bend down low during a walk through the city, there is a good chance we will be greeted by cute scalloped leaves and pretty purple flowers.

Though you can find a variegated cultivar sold at nurseries, ground ivy was not brought to North America for its pretty purple flowers. It was brought by English folks as part of a standard stock of household herbs. There is record of ground ivy being used medicinally in Europe, dating back to the first century. Numerous uses include treating bruises, coughs, fevers, headaches, jaundice, kidney disease, inflammation of the eyes, tinnitus, bronchitis, and asthma. That's quite a lot of healing power for this tiny plant.

Ground ivy was used in the fermentation of beer before hops in parts of Europe; in England, it is said that ground ivy was used for this purpose through the early 1500s. This legacy is remembered in an old common name, alehofe. Hofe is an olde English term for brewing. It was part of the local diet too, as well as being added to soups, stews, and salads for the minty/peppery taste. Upon introduction to this land, the Cherokee incorporated it into their healing regime to treat colds and babies' hives.

Long-tongued bees are always happy to see ground ivy, and in the past, it was a happy sight for humans too. There are tales from Greco-Roman mythology that ground ivy can cure melancholy. Sipping a hot cup of fresh ground ivy tea while looking out on the colors of spring sure sounds like an antidote to sadness to me.

Hairy Bittercress

Cardamine hirsuta

A winter or summer annual of the Brassicaceae family, the Mustard family. Both canola oil and mustard oil are obtained from plants in this family.

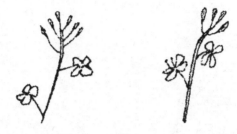

Habitat: Grows best in damp soil in either sun or shade but can tolerate dry, sunny sites as well. Can be found in sidewalk cracks, landscaped areas, gravel pathways, and lawns.

Came from Eurasia. Unknown how it traveled here—maybe hidden among ornamental flowers?

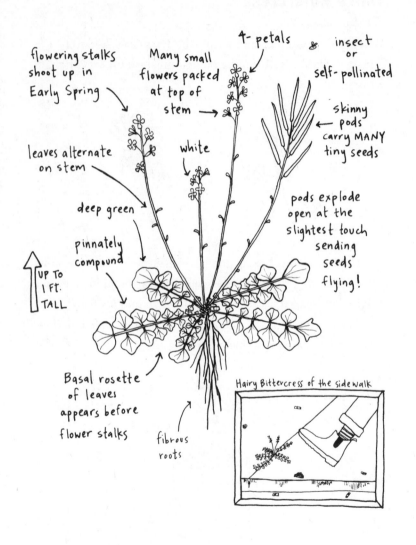

Hairy bittercress is full of hair. Well, not really. It actually doesn't seem hairy at first glance, but whoever named this plant sure thought so. The species name, *hirsuta*, means "hairy" in Latin.

It is not known how or when hairy bittercress arrived here, but when it did, it was united for the first time with some of its cousins. There are several species of the genus *Cardamine* (Bittercress) that are native to regions across North America. We may be less familiar with these native species because unfortunately, some are listed as threatened due to habitat loss. Hairy bittercress proves to be the only Bittercress that is able to survive the tough living conditions of the North American city. In fact, it has spread widely worldwide, capitalizing on disturbance by being quick to sprout on cleared land.

Seeds germinate in both the spring and fall. Fall germinated seeds overwinter as a basal rosette and send up the flower stalk at the first hint of warmer weather. One hardworking plant can produce a dozen flower stalks and hundreds of seeds. When touched or brushed by a gust of wind, the seeds burst explosively from the seed pods, sending them flying far and wide. Though its flowers are not as showy as some of the native Bittercress, it has definitely proven to be more adaptable. We urban dwellers benefit from its adaptability. The leaves can be eaten in spring and have a slightly peppery taste, not bitter at all. They are full of nutrients and, as a fresh green, are a welcome addition to the diet after we've been eating cellar root veggies all winter long.

Hedge Bindweed
Calystegia sepium

A perennial of the Convolvulaceae family, the Morning Glory family. Cousins to the annual morning glory (hailing from Central America) and the delicious sweet potato.

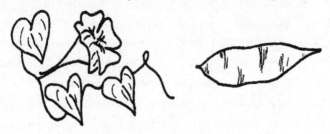

Habitat: Prefers moist soil, but is not necessarily selective. You can find it growing on chain link fences, hedges, roadsides, rock walls, and landscaped areas.

Comes from Eastern North America and Europe.

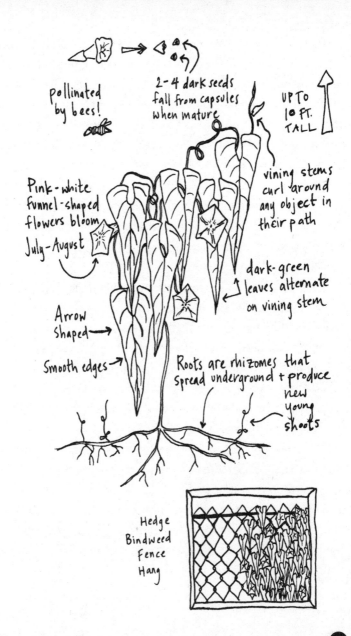

Sepium, the species name, comes from the Latin word *sepes*, which means "hedge." Is this name a warning to hedges that they will get covered if hedge bindweed gets close? Or is it a foreshadowing of what will appear when hedge bindweed covers a fence? Who knows? What I do know is that hedge bindweed can cover both hedges and fences—and fast. It can grow as much as nine feet in one year.

If you try to cut it down, it will come back the next year. Digging them up is your best shot at removal, but roots can reach up to ten feet underground and hedge bindweed can resprout from the tiniest fragment left in the soil. It is tenaciously tethered to life.

Many a gardener has let out a groan when they recognized hedge bindweed in the garden. Not the long-tongued bees though. They happily drink the sweet nectar and pollinate, ensuring the seeds and the spread of next year's plants. Hedge bindweed does not offer much medicinally, and it's not a go-to forage crop, though the stalks and roots can be cooked and eaten. It can aid in erosion control, which is super helpful on compacted, hilly sections of the city.

It may seem like hedge bindweed is straight nuisance. But the arrow-shaped leaves are pretty cool and the flowers are beautiful. And if that doesn't do it for you, then let it be, as us gardeners used to joke, your job security. Year after year, you know your purpose in the garden is to keep the hedge bindweed at bay. And what a wonderful thing it is: to know your purpose.

Horseweed

Conyza canadensis

An annual of the Asteraceae family, the Sunflower family. A very large family with many recognizable plants, there are twelve listed in this book!

Habitat: Thrives in dry soil in full sun. Can be found in vacant lots, public parks, median strips, highway embankments, railroad tracks, rubble dumps, and landscaped areas.

Comes from North America.

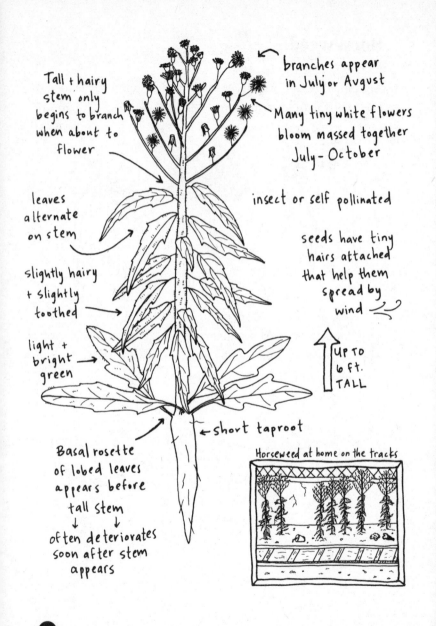

Horseweed is not a looker. It is tall, imposing, and hairy. The flowers are not showy or particularly pretty, and the branching flower stalks look crowded and messy. Can we look past that and see what else is there?

For one thing, horseweed is helping to keep taxonomists employed. You would think by now their job would be complete, but after much deliberation, taxonomists decided that horseweed is actually part of the genus *Conyza*, having formerly been considered part of the genus *Erigeron*, the same genus as daisy fleabane. Similar to daisy fleabane, horseweed was burned to ward off insects. Though that's a fun characteristic they share, it was not enough to keep them in the same genus.

However it is classified, horseweed has had a long history of use by humans. Traditional Native American medicines include an infusion of the roots taken to relieve diarrhea and stomach pains, a poultice of leaves applied to body pains, and a snuff made to stimulate sneezing during a cold. Horseweed was also boiled as a steam in sweat baths.

Upon the arrival of the American lawn, we forgot of our deep relationship with horseweed and began to target it as a weed with our chemicals. Horseweed has withstood an assault of herbicides for some time because of its amazing ability to remain resistant to them. Of the chemicals sprayed, it shuttles 85 percent of the glyphosate into its vacuoles where it can't damage the cell and the other 15 percent goes to leaf tips, making the individual leaf die back but leaving more than enough energy to survive. Wow.

Horseweed has been introduced and now spreads throughout Europe and Asia, so neither herbicides nor oceans can keep Horseweed down. You can see it blooming from afar: it stands tall and soft against the hard angles of the built city. No matter how we alter the landscape, Mother Nature sends her toughest allies to represent.

Japanese Knotweed

Polygonum cuspidatum

A perennial of the Polygonaceae family, the Buckwheat family, whose members include delicious buckwheat, rhubarb, and sorrel.

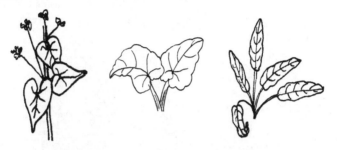

Habitat: Grows best on moist soil in full sun but can tolerate drought and shade. Can be found in vacant lots, along railroad tracks, highway banks, drainage ditches, and waterways.

Came from temperate East Asia. Intentionally brought to the US and sold as an ornamental plant in the 1870s. It has since escaped the garden and is listed as an invasive species in many states.

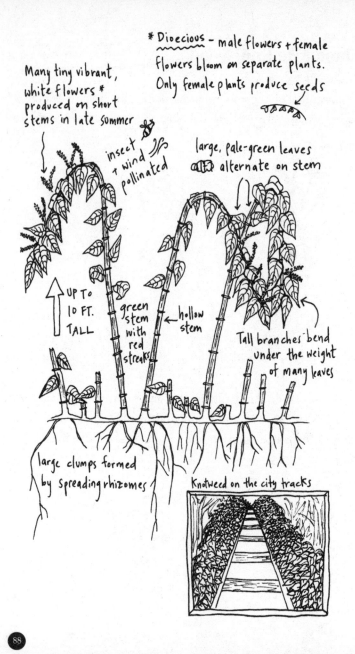

Japanese knotweed is infamous, known for its zest for life expressed by its vigorous spread and indestructibility. It can spread by more than twenty feet per year. How? The root system extends six feet underground and can swell and suffocate the root systems of other plants, creating plenty of room for the rhizomes of Japanese Knotweed to spread out and continue producing new shoots. Those new spring shoots are strong enough to break through asphalt and then can grow to a height of five feet in four weeks. The vigorous new growth produces plenty of leaves that photosynthesize energy from the sun to provide Japanese knotweed with needed strength to keep moving.

Japanese knotweed can resprout from even the smallest root fragment left in the ground or compost. These survival traits were developed in its native land where it is one of the first species to colonize land after lava flow. Plants must break through hardened lava and be able to withstand extreme levels of acidity and mineral pollution. It was bred to be tough.

Japanese knotweed is willing to share its love of life with us; the roots contain the antioxidant compound resveratrol, thought to promote longevity. And as we contend with Lyme disease against a growing tick population, Japanese knotweed root tincture is a helpful ally as a folk remedy. The young leaves can be cooked and eaten as a vegetable, and we can procure a yellow dye from its roots. It feeds a massive amount of pollinators, and permaculture expert Tao Orion suggests that we let cut stalks of Japanese knotweed do the work that beavers once did in stream channels.

It is unfair of us to point the finger at Japanese knotweed for pushing out native plants and changing the ecosystem when we are the ones who created the conditions for it. Let Japanese knotweed hold up the mirror so that we can see that we must change our way of life first if we want continued ecosystem degradation to stop. And as we remove it in order to heal our places, may it heal our bodies and minds.

Jewelweed

Impatiens capensis

A summer annual of the Balsaminaceae family, the Touch-Me-Not family, a small plant family of which the popular annual New Guinea impatiens is a member.

Habitat: Enjoys moist soil in either sun or shade. Can be found in woodland understory of public parks and along waterways and landscaped areas.

Comes from Eastern North America.

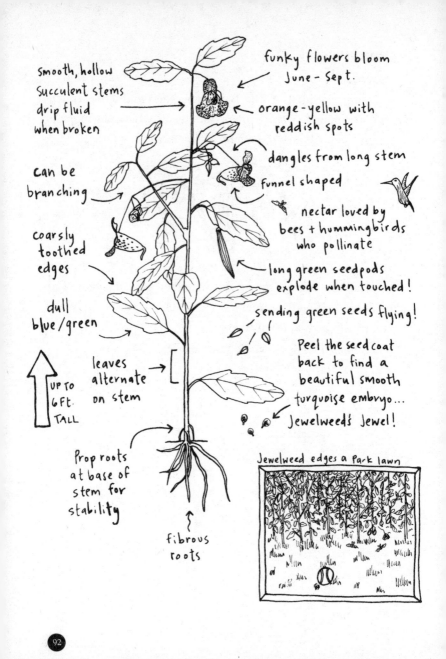

I remember the first time I met jewelweed as a young garden intern. I was gathered at the New York Botanical Garden with other NYC gardeners for an event of some kind. On a plant scavenger hunt, though we were not looking to find it, a teammate pointed out jewelweed. I was struck by the unusual shape of the flowers and how they could have so easily gotten lost in the mass of soft, light green foliage if it weren't for their brazen colors. Jewelweed's distinctiveness made it easy for me to identify and remember it: it's the only plant I can recall from that scavenger hunt.

Plants like jewelweed make the task of identifying much less daunting. Start with plants that you find easy to identify and get to know them. Visit them often, search for them, and soon, they will introduce you to plants nearby. Slowly your confidence and plant knowledge will grow, just like jewelweed does, every year.

The fantastical ejection of seeds by the seed pods ensures that jewelweed does indeed grow every year. At the slightest touch or gust of wind, they explode open, sending seeds flying! Jewelweed is often found in large clusters, each ejected seed landing spaced apart to grow unencumbered.

Jewelweed is prized for a poultice made from its leaves that can relieve the pain and itch from bee stings, nettle stings, and poison ivy. And the whole plant can produce a yellow/orange dye. Some say that the way beads of water from rain or dew form sparkling droplets on the leaves is how this plant became known as jewelweed and not because of the turquoise embryo. Either way, this plant is a gem.

Lambsquarters

Chenopodium album

Lambsquarters is an annual plant of the Chenopodiaceae family, the Goosefoot family. Relative to the following well-known food crops: spinach, quinoa, Swiss chard and beets.

Habitat: This family is known to love dry conditions and for being very salt tolerant. Found in many urban habitats, including building foundations, sidewalk cracks, tree pits, rubble piles, railroad tracks, landscaped areas, and median strips.

Came from Europe. Maybe by way of someone's shoe? Once here, it traveled far and can now be found in every state. I mean, a large plant can produce up to 75,000 seeds—that's a lot of babies!

Lambsquarters can now be spotted at your local farmer's market sold in bunches amongst the cultivated crops. We Americans are finally catching on to what our ancestors knew: this plant is packed with vitamins. Europeans ate it as a potherb, and there is record of at least twenty-seven Native tribes using the cooked greens or seeds in their diet. Lambsquarters did not get dismissed in India and Pakistan; in the Punjab region, lambsquarters is cultivated as an important food crop!

The species name, *album*, comes from the Latin word *albus* meaning "white." The leaves are covered in many tiny, wooly hairs that give their undersides a powdery, whitish appearance. *Chenopodium* is Greek for "goosefoot," a nod to this plant's leaf shape.

Lambsquarters has found a home in the city thanks, in part, to a trait called morphological plasticity. This trait allows it to adapt to the environmental conditions in which it grows; in rich soil, it is huge and happy, and in poor soil, small and sad. And, amazingly, in both cases, it sets seed. Because of morphological plasticity, lambsquarters is a good indicator of soil fertility.

Lambsquarters' long taproot draws up water and nutrients from deep in the soil, and what it doesn't use is made available to others with more shallow root systems. Birds love the seeds, and many insects feast on the leaves. The seeds have been successfully germinated after being found in an archaeological site dated as being 1,700 years old! This plant has real staying power and ensures a high nutrient food source for the future generations of many species. Its presence is a soft and gentle reminder that we are all here for each other.

Mallow

Malva neglecta

An annual of the Malvaceae family, the Mallow family. Did you know that okra and cotton are relatives? As well as hollyhock and rose-of-sharon!

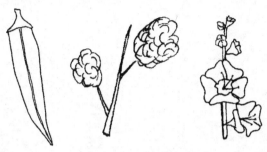

Habitat: Grows best in full sun and rich soil. Can be found in vacant lots, public parks, landscaped areas, roadsides, and walkways.

Comes from Eurasia. Unknown how it traveled here, though it is believed to have been introduced by the first arriving Europeans as a food source.

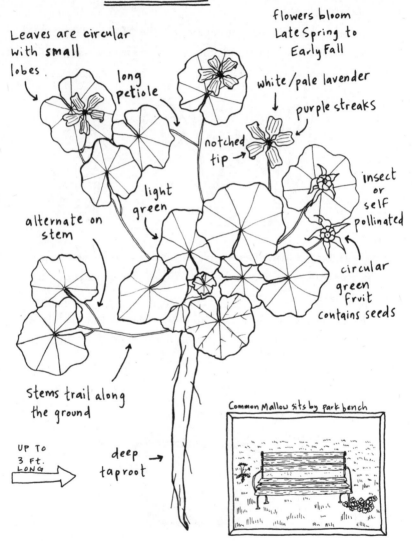

Neglecta, the species name, comes from the Latin word for "neglected," and it speaks of the land on which common mallow grows. It's a good thing we leave soil at the surface, untouched by paving, for common mallow to thrive, because it is such a beautiful plant that supports a diverse urban ecosystem. The flowers are food for insects and the leaves are a larval host for a number of species of moths and butterflies.

Though it thrives in a neglected landscape, common mallow itself has not been overlooked by humans. Historically in Europe, it has been used for its demulcent, laxative, and anti-inflammatory properties. The leaves and young shoots can be eaten, and the mucilaginous quality of the leaves can be used to thicken soups, much like common mallow's cousin okra. An egg-white substitute for meringue can be made by simmering the roots in water, then when the water thickens it can be whisked to create the meringue. Now, that is a handy thing to know.

It is called cheeseweed in some circles because of its cheesy-looking fruits that are enjoyed by snacking children. No matter what you call this plant, common mallow, cheeseweed, or something else, it is a welcome addition to any neglected land. The fibrous roots hold the soil in place, and the flowers, fruits, and leaves bring a multitude of beings together to marvel and feast. Suddenly, the land does not seem as neglected.

Milkweed

Asclepias syriaca

An herbaceous perennial of the Asclepiadoideae family, the Milkweed family. Like some of its cousins, it is a host plant for the larvae of the monarch butterfly.

Habitat: Sunny, disturbed sites with a wide tolerance of pH range. Tolerates low-nutrient soil. Can be found in median strips, vacant lots, public parks, landscaped areas, railroad rights-of-way, gravel paths, and rubble dumps.

Native to Eastern North America.

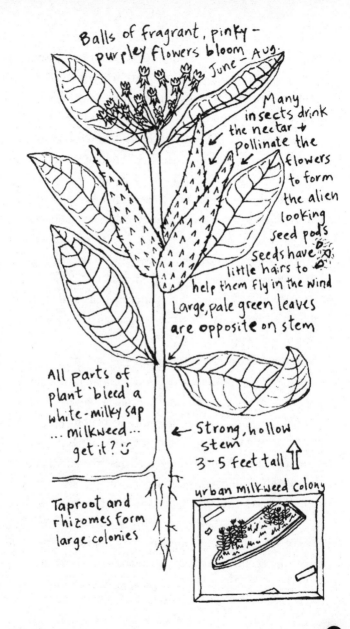

Milkweed's botanical name, *Asclepias*, is in honor of the Greek god of medicine, Asklepios. A testament to our deep relationship with this medicinal plant. IMPORTANT NOTE: The active chemical in milkweed, the very same chemical that is the source of its medicine, can be toxic in large amounts.

There are records of many traditional uses of milkweed by Native tribes of the Eastern US: the seed pods were used as soup thickeners or as a vegetable similar to okra and young shoots and flower buds eaten as vegetables. Medicinally, milkweed leaves were traditionally taken as a laxative and urinary aid and the sap applied externally to remove warts. Dried stalks were harvested for sewing thread, fishing lines, and bowstrings. There is an anthropological record of a forty-foot-wide deer net that contained around 7,000 feet of milkweed cordage. Wow, that is an amazing amount of milkweed and people power.

Milkweed produces a significant amount of nectar which feeds many insect pollinators. Once pollinated, the seeds are formed in alien-shaped pods and are released to fly in the wind when mature. It is a beautiful sight to behold and also the source of another common name for this plant: cottonweed!

Though it is moving quickly across North America due to deforestation and considered invasive in some parts of Europe, milkweed is still being sold at nurseries, promoted as a plant for the pollinator garden. And in the urban environment, milkweed supports a diversity of life that is essential for keeping the ecosystem thriving. Its sturdy stem says that just like our built skyscrapers, nature too stands tall.

Mugwort

Artemisia vulgaris

Mugwort is a perennial of the Asteraceae family.

Related to wild quinine, dahlia, yarrow, dandelion, and many, many, many others. It is part of one of the largest plant families, after all.

Habitat: Tolerant of disturbed, compacted soil with low pH (acidic). Salt tolerant. Found in public parks, sidewalk cracks, vacant lots, rubble piles, construction sites, railroad tracks, highway embankments, median strips, fence lines, and building foundations.

Came from Eurasia, likely introduced to North America for medicinal purposes.

It has since spread throughout the continent.

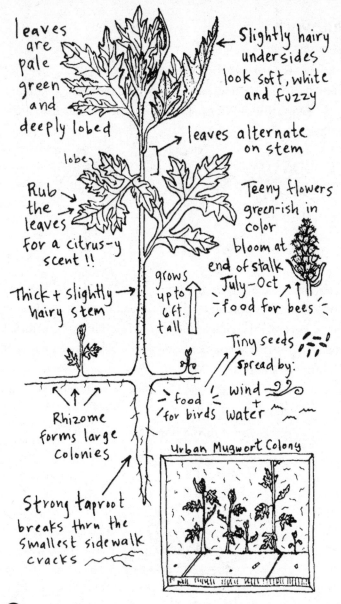

The genus *Artemisia* is named for Artemis, Greek goddess of the hunt, wild nature, and childbirth. This is a large genus containing many fragrant, edible, and medicinal herbs, including tarragon and wormwood (of absinthe fame).

The species name, *vulgaris*, means "common" in Latin and attests to the fact that once you know mugwort, you see it everywhere. Mugwort has spread across Europe, Asia, Africa, and North and South America where people have both welcomed it for its amazing medicinal qualities and shunned it for its tenacious tendencies.

Its medicinal qualities have long been prized in Eurasia. In Europe during the middle ages, it was known as the mother of all herbs. It is immortalized in the Nine Sacred Herbs charm by the Anglo-Saxons. Mugwort was once used to flavor beer and is a popular herb for lucid dreaming. Wort means "herb" in Old English, a name to help the people share the herbal wisdom that this plant contains. In China, mugwort continues to be used by acupuncturists in moxa, to treat colds, and to turn a breech baby. Mugwort is grown in India as well, prized for its use in Ayurvedic medicine.

Mugwort and humans have a long history together and the story continues on the city streets. The poster child for urban plant life, mugwort is able to withstand all kinds of tough living conditions. It can regenerate when cut, and even the smallest piece of root left in the soil can come to life again. Mugwort gives back to its home. Its use in phytoremediation efforts make the land less toxic, a benefit to all layers of the urban ecosystem. Removing cadmium, copper, and nickel by extraction is its specialty.

The muted, sweet scent of the bruised leaves brings a sense of calm in the midst of the hustle of the city. And the rocking of the silvery-green foliage in the breeze can transfix even the most busy body.

Mullein

Verbascum thapsus

An evergreen biennial of the Scrophulariaceae family, the Figwort family. Related to some much loved garden plants: snapdragon, foxglove, veronica, and penstemon.

Habitat: Loves dry, sandy, disturbed soil. Vacant lots, gravel paths, sidewalk cracks, building foundations, railroad tracks, roadsides, and median strips.

Came from Eurasia. Likely introduced for medicinal purposes in the mid-1700s. Has since spread from coast to coast.

When mullein is in its flowering year, it cannot be missed. The flowering stalk, tall and thick, lights up above its surroundings. Perhaps this is why early Europeans first thought to dip the dried flower stalks in wax and burn it for use as a torch, lending mullein another common name: candlewick. The pollinators don't miss mullein either: bees, wasps, flies, birds, you name it—all come to drink its sweet nectar.

It was a prized medicinal herb in its native land, a well-known healer of the respiratory system. A tea of mullein leaves can be used to treat colds, bronchitis, and asthma, and mullein oil is helpful for removing excess earwax and tending ear aches. When mullein was introduced to North America, people on this land were quick to incorporate its medicine into their traditions. The Abenaki made a necklace of the root for teething babies!

It was probably brought to North America intentionally, but I bet even if it hadn't been, mullein would have found its way here. One plant can produce up to 180,000 seeds. Seeds are shaken from the stalk when the wind blows so they do not travel far, but over time and generations, they can really cover a lot of ground. In Europe, seeds that had been buried for close to five hundred years germinated when brought to the surface of the soil. Mullein is full of surprises and here to stay.

Being in the presence of mullein feels like a spiritual experience. Picturing the flowering stalk as a flaming torch makes me wistful for a time when we lived more unified with nature. Mullein's torch can guide us back.

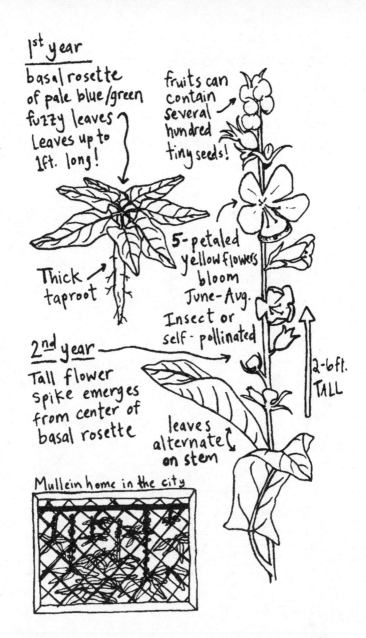

Pennsylvania Smartweed

Polygonum pensylvanicum

Pennsylvania smartweed is an annual of the Polygonaceae family, the Buckwheat family. Related to fellow urban plants Japanese knotweed and curly dock, as well as everyone's favorite pie filling, rhubarb!

Habitat: Loves full sun and rich, moist soil. Found in many urban habitats, including tree pits, sidewalk cracks, vacant lots, building foundations, edge-of-park ponds, and landscaped areas.

Native to Eastern North America. Found in fields, moist areas at the edge of the forest, and other moist waste places.

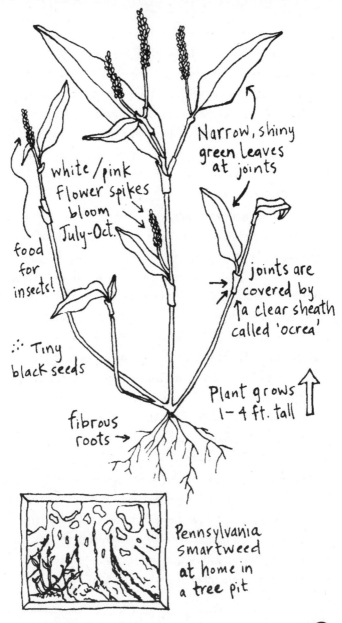

Pennsylvania smartweed is named for the state in which it was found. A name that was given to the land by William Penn, an Englishman who founded the Province of Pennsylvania. Before William Penn, the land and the plants that inhabited it had other names, named by the people who, for generations, called that land home. I was unable to find a name for Pennsylvania smartweed in a language other than English, but there is record of the relationship between the plant and its people.

Pennsylvania smartweed has been traditionally used for its antihemorrhagic, anticonvulsant, and antihemorrhoid properties by the Chippewa, Menominee, Iroquois, and Meskwaki tribes inhabiting current day Wisconsin, Minnesota, Michigan, North Dakota, Ontario, New York, and Iowa. The plant was also used for gynecological issues.

Many species of beetles, flies, and bees enjoy the nectar and do the pollinating. Seeds are an important food source for songbirds and waterfowl. Many migrating songbirds pass through cities on their journey; staying well-fed is important for them to reach their final destination and ultimately, for their survival. Plants and animals, working together to ensure future generations.

Pennsylvania smartweed seeds sprout readily in sunny, disturbed sites. The compacted soil it finds in the city is low in oxygen, just like the moist soil of its native habitat, showcasing the adaptability of Pennsylvania smartweed. The pretty pink flower spikes that indiscriminately beautify call to us city folk to remember the way this land used to be.

Pineapple Weed

Matricaria discoidea

A summer annual of the Asteraceae family, the Sunflower family. Many, many relatives, such as white wood aster, aromatic aster, Tatarian aster, and white snakeroot.

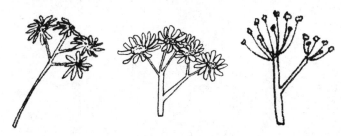

Habitat: Likes full sun and is an indicator of sandy or compacted soils. Can be found in tree pits, sidewalk cracks, rubble dumps, vacant lots, gravel pathways, and railroad tracks.

Comes from Western North America and has since naturalized across the continent. Probably hitched a ride. . .

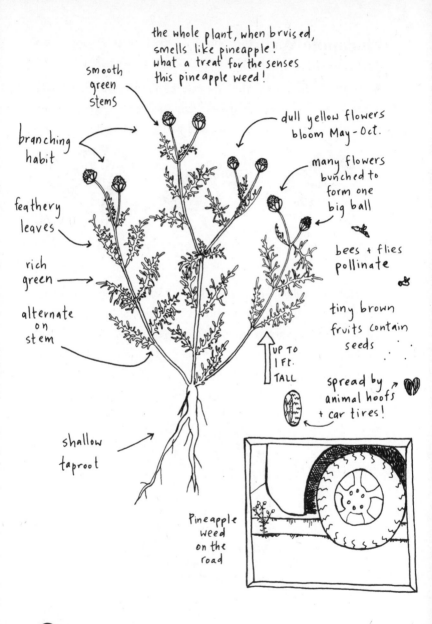

How pineapple weed got its name is apparent to me every time I bend down to admire it. The shape and color of the flowers look just like mini pineapples, and even if you don't see it, try to tell me that the fragrance of the crushed leaves doesn't transport you to a tropical paradise covered in pineapples—just try. The scent comes from a compound called myrcene, an important compound in the fragrance industry that is also found in lemongrass, thyme, and cardamom.

Native tribes of the West Coast have long enjoyed the scent, using pineapple weed in sweat lodges and steam baths for its pleasant odor. The plant can be dried and ground for use as a perfume, and the sweet flowers are sweet treats for children. Pineapple weed was also used medicinally for stomach pains associated with gas, diarrhea, and general upset stomach, ailments for which we now turn to its look-a-like, chamomile.

Pineapple weed is one of the twentieth century's fastest spreading plants thanks, in large part, to the advent of cars. Its seeds evolved traits to help it spread by animal hooves, which, turns out, makes them perfectly suited to be spread by the automobile. Like no other plant, pineapple weed has been able to take advantage of car tires, lodging in the treads for a fast and far transport.

The sweet pineapple weed barely takes up any space. The flowers and foliage of this tiny plant can transport us to a tropical paradise faster than any car. Look for it as a place of tranquility in the city, as you would look for an oasis in the desert.

Pokeweed

Phytolacca americana

A perennial of the Phytolaccaceae family, the Pokeweed family, a small plant family that contains less than one hundred species.

Habitat: Grows best in moist soils in full sun. Can be found in vacant lots, building foundations, public parks, and railroad tracks, as well as along the waterfront.

Native to North America.

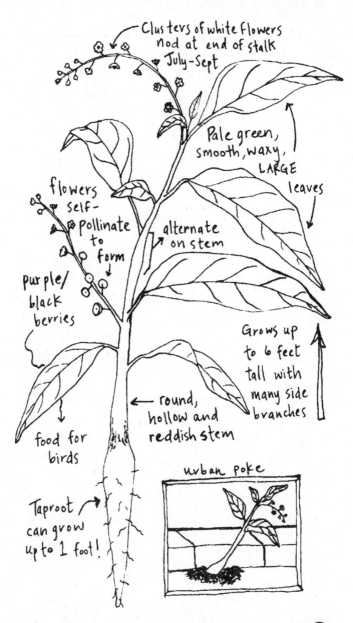

Pokeweed is such a vibrant, tropical-looking plant, it's almost hard to believe that it was made of this place! The large branching form and bright greens, pinks, and purples were quite conspicuous in its original habitat, the forest edge, and remain conspicuous in its urban home.

The attention-seeking pokeweed is calling out to songbirds that food is available. Birds love the berries, which remain an important food source. Some mammals eat the berries too, including the urban dwelling raccoon. Certain species of flies and bees visit for the nectar. Pokeweed solidifies its place in the ecosystem by providing food and in return, the next generation is transported to a new home. To give and receive is a joy of life.

The common name is said to be derived from the Algonquin name of this plant, pocan, which means "bloody" and alludes to its use as a dye plant. The berries can be made into a pale pink dye or ink. The medicinal uses of pokeweed by several Native tribes of the East include as a poultice for ulcers and swelling and an infusion of berries for arthritis. The whole plant is toxic and should not be eaten raw, though the young shoots have been eaten for centuries, especially in the Southern US after being boiled in several batches of water, a dish known as poke sallet.

Pokeweed can regenerate from the large taproot when cut down and its seeds remain alive in the soil for many years—two traits that have helped pokeweed survive despite the change of scenery and its weedy status. The colors and the arching branches are just begging to be immortalized on canvas. A true work of art, pokeweed graces our cities with its presence, and we are lucky to have it.

Prickly Lettuce

Lactuca serriola

An annual or biennial of the Asteraceae family, the (very large) Sunflower family. Related to the popular ornamentals Joe Pye weed, helenium, and prairie dock.

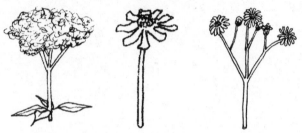

Habitat: Loves full sun and nutrient-rich soils but can tolerate dry sites and poor soils. Can be found in vacant lots, median strips, rubble dumps, railroad rights-of-way, building foundations, sidewalk cracks, and public parks.

Came from Europe. It is unknown how it traveled here, though probably introduced to North America in the late 1800s as a contaminant in seed. Has since spread across the US.

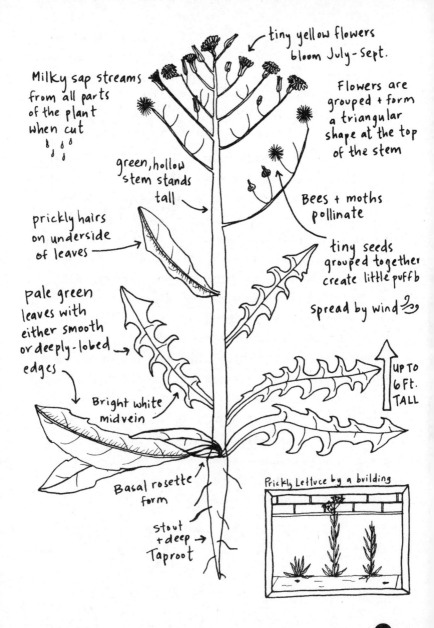

Prickly lettuce is a mystical plant. There is a long history of its magical uses in the UK and across the Mediterranean. In ancient Egypt, it was used primarily as an aphrodisiac, dreaming herb, and calming nervine. Have you ever sat with prickly lettuce? You can feel the uplifting energy of possibility.

It is also the wild ancestor of lettuce, the well-loved foundation of salads! The botanical name for lettuce is *Lactuca sativa*; *sativa* means "cultivated." It is best to stick with eating the cultivated variety as prickly lettuce leaves contain potent alkaloids which can be toxic in high amounts. The milky sap that exudes from the whole plant contains lactucin and lactucopicrin, two compounds that are calming to the central nervous system and help relieve pain. The dried milky sap was included in the US pharmacopeia from 1820–1926. Science is backing up what our ancestors knew. In some countries, prickly lettuce sap is used as a nonaddictive alternative to opium poppy sap. The sap can be dried and smoked and deliver mild euphoric effects, which is how it got another common name, poor man's opium.

It is so cool that a plant of this nature grows unassuming in our cities. These plants have secrets that they are just dying to share with us. All we have to do is slow down and listen. Prickly lettuce is here to help us with that.

Purslane

Portulaca oleracea

A summer annual of the Portulacaceae family, the Purslane family. A small family.

Habitat: Loves nutrient-rich sandy soil in full sun but can tolerate compaction, drought, and salt. Can be found in sidewalk cracks, building foundations, vacant lots, public parks, landscaped areas, and rubble dumps.

Comes from Eurasia and lives in MANY places throughout the world.

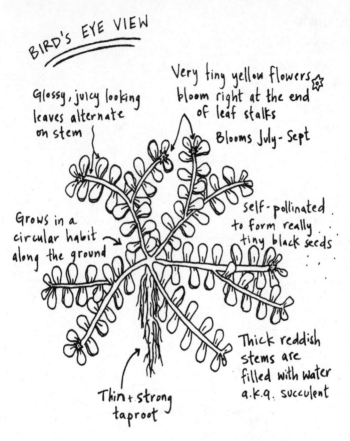

Purslane is an amazing little plant that has lived beside humans for generations, and I mean many generations. The species name, *oleracea*, means "of the vegetable garden," proof that purslane has accompanied us for a long time. Because of its global presence and long history of use throughout the world, it is unclear where it originated, though its succulent stems suggest an adaptation to the deserts of North Africa, the Middle East, and the Indian subcontinent. It is also unclear when purslane was introduced to North America. It definitely could have come with Europeans in the 1700s, brought purposefully as a salad green and soup thickener, but there is archaeological evidence that purslane was growing in North America long before then.

The leaves are a rich source of Omega-3 fatty acids, high in Vitamin A and delicious. People all over the world have made use of purslane as a healing herb—in Palestine it is used for renal failure; in Nigeria as a fertility aid; in Sierra Leone to treat hernias, stop bleeding, and induce abortion; in Democratic Republic of Congo to treat gonorrhea; in India to treat excessive mucus, cough, and diarrhea; and in traditional Chinese medicine it is called "the vegetable for long life" and used for a variety of ailments.

It makes me so sad to think of all the purslane that I've sent to the compost pile, or worse, the landfill, because of what I did not know as I tended my gardens. I cannot unknow what I know now and I can no longer see a weed. I readily remove purslane when I find it, already dreaming of the dishes I'll make. I do not always take. There are times I let it live and leave it in reverence for this plant that has traveled so far and seen so much.

Quackgrass

Elymus repens

A perennial of the Poaceae family, the grass family. Related to the crops that bring us a variety of alcoholic beverages, such as sake, whiskey, bourbon, and rum.

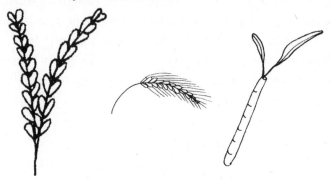

Habitat: Loves full sun and good soil. Can be found in vacant lots, sidewalk cracks, fence lines, building foundations, median strips, and highway embankments.

Comes from Eurasia and North America.

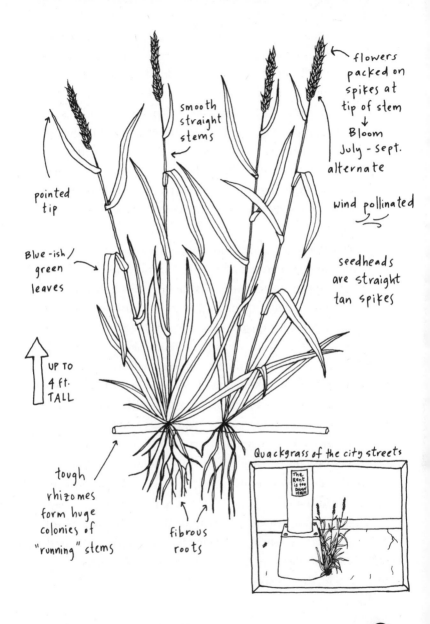

The common name quackgrass is actually a corruption of an earlier used name, quickgrass. It was named quickgrass because of its ability to green up quickly in Spring. For that reason, it was commonly planted as a forage crop in the nineteenth century. How times have changed. Quickgrass is now quackgrass, and quackgrass is now an agricultural weed, hated for the same reasons it was once loved. Doesn't that tend to happen in relationships: the traits we first fell in love with become what aggravates us most a few years in? Seems like humans and quackgrass are in need of some relationship counseling.

Quackgrass is deemed too competitive because the root system is huge, is dense, and crowds out other plants. Then, the sharp-tipped rhizomes can puncture through the root balls of other plants, quickly crowding them out. Besides being the vehicle for quackgrass' spread, the rhizomes store food for the grass, which has been food for us in times of scarcity, ground up into a flour and mixed with wheat to make bread. Quackgrass accumulates high amounts of silica and potassium, important minerals for soil health, so next time you are pulling quackgrass, save the roots to make an organic fertilizer by soaking in water for a month or two. The roots of quackgrass have a dual nature, just like the rest of us.

In the urban environment, quackgrass never goes too far in its spread thanks to concrete. Where it does go, its foliage provides food for insects and its seeds food for birds. It is still quick to green in spring, a welcome sight for our grayed-out eyes. Another tough city competitor, quackgrass helps to maintain the balance between human-made and Mother Earth.

Ragweed

Ambrosia artemisiifolia

A summer annual of the Asteraceae family, the Sunflower family. Related to tarragon, wormwood, coltsfoot, and the rest.

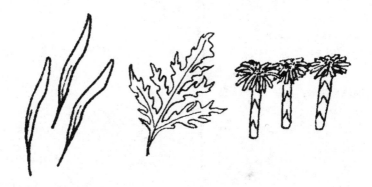

Habitat: Does not discriminate between soil type and can do well in moist or dry conditions. Can tolerate high pH and salt. Can be found in vacant lots, sidewalk cracks, building foundations, railroad tracks, rubble dumps, median strips, and landscaped areas.

Comes from North America.

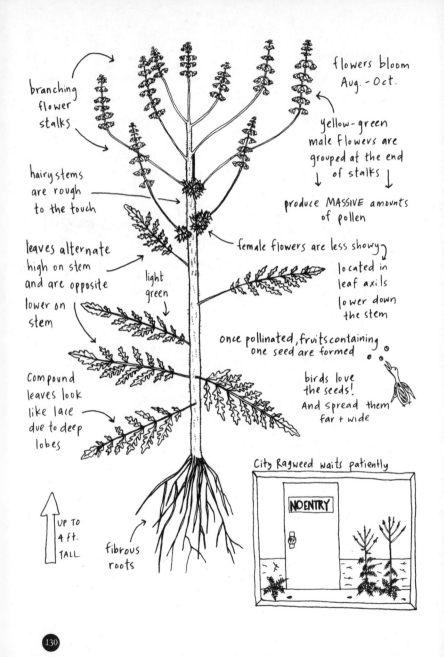

The genus name, *Ambrosia*, means "food of the gods." Clearly Ragweed held a special place in the hearts of those who named it. The species name *artemisiifolia* tells us that while ragweed is most definitely of the *Ambrosia* clan, its leaves are similar to those in the genus *Artemisia*. Once you learn to recognize those leaves, you will see them everywhere. Ragweed is common throughout North America and has spread and is considered invasive in parts of Central and South America, Europe, Asia, Africa, and Australia.

The massive amount of pollen produced (the cause of hayfever for folks in late summer and fall) is carried by wind and ensures that flowers get pollinated and seeds of the next generation are formed. Those seeds can remain viable for up to eighty years, waiting for the optimal conditions to sprout. This, coupled with ragweed's high tolerance for many varied conditions, helped it travel and flourish in urban centers around the world.

To the dismay of allergy sufferers the world over, ragweed has been found to produce more pollen in air with higher levels of carbon dioxide. Good thing then that the pollen is being harvested and used commercially in pharmaceuticals to counteract, of all things, hayfever! Ragweed is also useful in phytoremediation efforts, especially useful in removing lead from the soil. Helping us and the earth heal.

There is evidence that some ragweed species were cultivated in Native settlements for their protein-rich, high fat grain. There are 220 insects that are known to enjoy its foliage, and the seeds are an important food source for migratory birds and resident birds alike, as well as chipmunks and voles. After consideration, it seems this plant does live up to its name, a food of the gods indeed.

Red Clover

Trifolium pratense

A perennial of the Fabaceae family, the Pea family. Related to many well-known food crops and ornamentals: peanuts, lentils, chickpeas, redbud tree, soybeans, alfalfa, wisteria, and the list goes on and on and on.

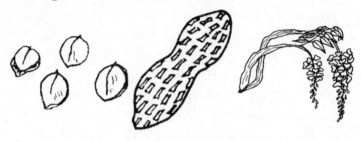

Habitat: Sunny, disturbed soils. Pairs with rhizobium bacteria to fix nitrogen in the soil, making it available for it and other plants so it can thrive in nutrient-depleted soils. Can be found in vacant lots, public parks, rubble dumps, and landscaped areas.

Came from Eurasia. Maybe as a stowaway in a ship's ballast?

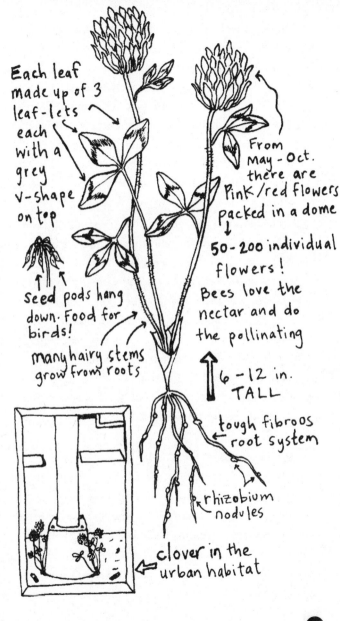

The botanical name of red clover is quite descriptive. The genus name, *Trifolium*, Latin for "three leaf," references the three leaflets of each leaf. *Pratense*, the species name, is Latin for "of a meadow," and that is exactly where red clover can be found in its native habitat. Red clover is named for the color of the flowers, which look more pink to me, but we don't need to get Pantone specific. It has made such an impression in its adopted country that Vermont claims it as its state flower. It really is a sight to behold, a sea of red clover blooming on fallow farmland.

As it blooms, red clover feeds so many. Its nectar is loved by bees, butterflies, and moths, and various species of caterpillars feed on the foliage. At the same time, red clover creates food for other plants by working with rhizobium bacteria to fix nitrogen in the soil. Artificial nitrogen fertilizer was created just over one hundred years ago and continues to be so damaging to our environment. It creates about 1 percent of all human emissions of carbon dioxide, and fertilizer runoff is a major pollutant to our waterways. Red clover and other members of the Fabaceae family can be and, arguably, should be used as cover crops or green mulch in its stead.

Besides helping farmland, red clover is a medicine for the urban environment. It can be used in phytoremediation efforts to remove zinc from the soil. Just as it heals the land, red clover heals our bodies. It is known as a healing herb for the reproductive system. Rich in calcium, phosphorus, and Vitamin C, the leaves and flowers can be added to salad or tea for a nourishing boost of all kinds. Red Clover gives us so much; the least we can do is give some r-e-s-p-e-c-t.

Redroot Pigweed

Amaranthus retroflexus

An annual of the Amaranthaceae family, the Amaranth family. Just like gomphrena and celosia, popular ornamental flowers.

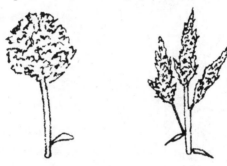

Habitat: Prefers moist, rich soil in full sun. Can be found in public parks, vacant lots, landscaped areas, median strips, and building foundations.

Came from Central America, unknown whether on purpose or not. Perhaps spread by a bird?

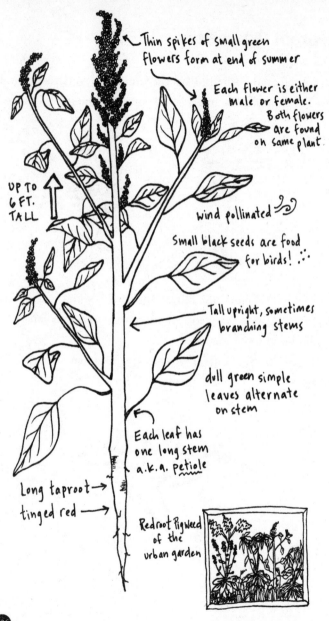

The genus name, *Amaranthus*, is Greek for "unfading" and refers to the everlasting flowers. The flowers do keep their color and beauty for a long time, even after seeds have matured, perhaps to attract the birds to their bounty. The seeds are an important food source for many species of birds.

Redroot pigweed is also food for humans. The seeds and leaves of the redroot pigweed and its cousins (other Amaranth species) were a major part of the local diet pre-colonization. The leaves were eaten as a potherb and the seeds were winnowed and ground into a flour; they still can be. There are claims that redroot pigweed can cause nitrate poisoning of livestock if overconsumed because it has a tendency to accumulate and hold nitrate in its leaves. These claims are used to justify its eradication. However, this is only an issue if livestock are grazing on lands where chemical fertilizers are used—another reason to question the need for these fertilizers and instead make use of the many Fabaceae plants (some of which are profiled in this book) as cover crops or green mulch!

Redroot pigweed is an immigrant making a home in the city. It carries the ways of being from the old country and does not want us to forget our roots. Our roots are in the land. And we need not worry that redroot pigweed will leave us; its deep taproot that draws water and nutrients up to the surface of city soil and its prolific seed production ensure that redroot pigweed is here to stay. Redroot pigweed is everlasting.

Reed

Phragmites australis

A perennial of the Poaceae family, the Grass family. Related to bamboo, lemongrass, rye, corn, and many other familiar species!

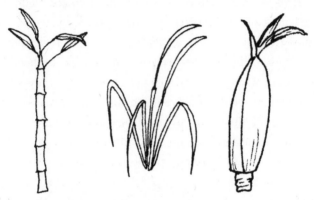

Habitat: Dominates sunny wetlands throughout the temperate world. Tolerant of salt. Can be found in vacant lots, landscaped areas, public parks, and drainage ditches.

Comes from Asia, Europe and North America.

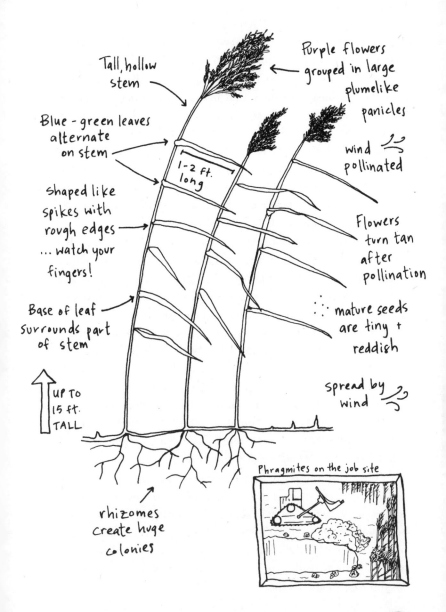

I remember when I first met *Phragmites*, that is what we called it at the public park. My fellow garden intern recognized it and was horrified that it had made its way into our garden. I became horrified too, and we both jumped in to remove the long rhizomes that had started to spread in one garden bed in the Northern corner of the park. It was a game. Who could pull the longest stretch of rhizome? *Phragmites* definitely won that day. We both tired of digging and pulling long before we removed the entire patch. We returned to that patch throughout the entire season, never feeling satisfied that we completely removed the *Phragmites* because even just a tiny root fragment left in the soil can resprout.

Its tenacious root system coupled with the massive amount of wind-blown seeds produced, and it is no wonder that *Phragmites*, or common reed, is one of the most widely distributed flowering plants, occurring on every continent except Antarctica. Humans have made good use of common reed where they've found it growing. Across the land of its spread, common reed stems have been (and in some cases still are) used as a source of paper, to thatch roofs, as quills for pens, for basket weaving and mat making. Hollow shafts turned into musical instruments, smoking pipes, and splints for broken bones. A newfound offering of common reed is cleaning bodies of water through phytoremediation efforts at waste treatment facilities. A noble task!

The winter after I met common reed, the time of the gardener's reflection, I thought, *What is the purpose for one species to hold on to life so strongly?* The answer at that moment was that it is possible to resprout from whatever source is left within, no matter the challenges or takedowns life throws our way. Common reed taught me the power of regeneration, a lesson that I carry today.

Shepherd's Purse

Capsella bursa-pastoris

An annual of the Brassicaceae family, the Mustard family! Just like many of our favorite food crops: kale, cabbage, broccoli, radish, mustard, cauliflower, and the list goes on.

Habitat: Can tolerate a range of growing conditions. Found in vacant lots, tree pits, sidewalk cracks, building foundations, stone walls, and landscaped areas.

Came from Europe shortly after colonization in the 1600s. Unknown whether on purpose or not. Maybe it came packed into packing materials? Can be found throughout the world.

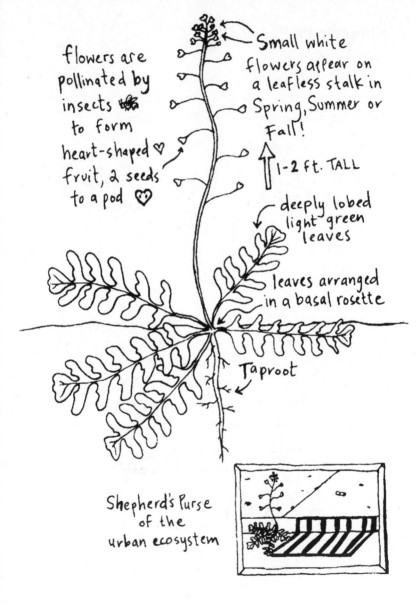

The common name as well as the species name refers to the shape of the seed pod, which looks like, you guessed it, a vintage shepherd's purse. This really makes me want to see an actual, old-timey shepherd's purse—they look so cute! It is quite recognizable in the field thanks to the seed pod, and that is a good thing, for it is really quite useful.

Like other members of the Brassicaceae family, shepherd's purse is rich in Vitamin K, so it has historically been used to treat issues of the blood, such as hemorrhages, menstrual issues, internal bleeding, and nosebleeds. It can be eaten and is even cultivated in China for consumption. I've never seen this myself, but apparently the seeds release a viscous compound when wet and aquatic insects, including mosquito larvae, get stuck to the seeds and eventually die. An amazing alternative to pesticides for controlling the mosquito population! Shepherd's purse is said to be the second most common weed worldwide, so if we all work together and throw some seeds in standing water near our homes, it may prove to be a pretty effective way of reducing mosquitos!

Seeds remain alive in the soil for many years and sprout readily upon disturbance, which is how this plant has become so prevalent in our cities. The cute flowers and funky pods inject a playfulness into the day, providing motivation and inspiration to live a full life.

Smooth Crabgrass
Digitaria ischaemum

A summer annual of the Poaceae family, the Grass family. Cousin to all true grasses, including all turf (or lawn) grass and ornamental grasses, such as switchgrass, feather grass, and prairie dropseed.

Habitat: Loves sun and can tolerate drought. Can be found in sidewalk cracks, gravel paths, building foundations, tree pits, public parks, trampled lawns, and landscaped areas.

Came from Europe. Unknown how it got here: perhaps brought by a cow?

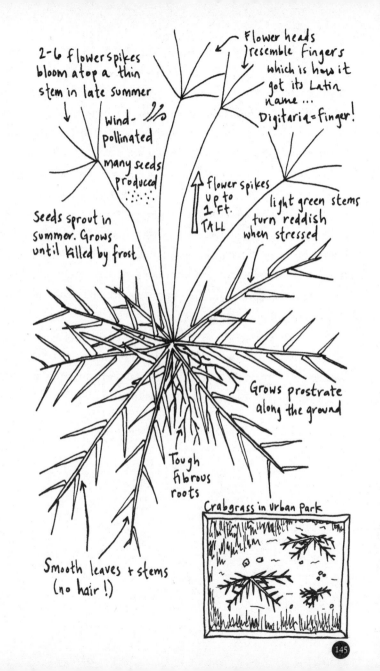

Like a lot of the plants in this book, once you know smooth crabgrass, you will see it everywhere. Especially in lawns of parks and homes. When turf grass suffers and dies from drought conditions, it is often smooth crabgrass that swoops in to fill in the blank spaces. As Aristotle is thought to have said, "Nature abhors a vacuum."

Smooth crabgrass was once used as a forage grass in the Southern US, where it thrived as a summer annual when other cool-season grasses failed. The caterpillars of some species of moths eat the foliage, as do other insects, and songbirds enjoy the seeds. Seeds have been spread across the continent thanks to human foot traffic, maintenance equipment, and seed contaminated soil, making the animals who feed on smooth crabgrass very happy.

Smooth crabgrass is no striking beauty, but consider it a carpet for the bare ground. It softens the spot for your blanket as you take a minute (or twenty) to breathe and be, ignoring the musts and the shoulds of city life.

Spotted Spurge

Chamaesyce maculata

An annual of the Euphorbiaceae family, the Spurge family. Related to the houseplants croton and rubber plant, as well as the food crops tapioca and cassava and the plant of laundromats and Christmas: poinsettia.

Habitat: Loves full sun. Tolerates dry soil, compaction, and foot traffic. Found in sidewalk cracks, building foundations, gravel pathways, railroad tracks, and landscaped areas

Comes from Eastern North America.

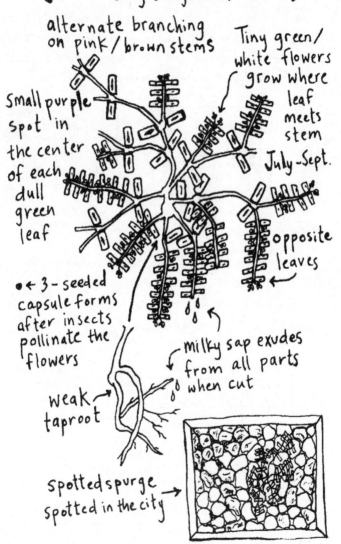

This diminutive plant is easy to rush past and over. It doesn't have the showy, colorful leaves of its relatives, or does it? If you bend down close, you'll be treated with a splash of purple on a green leaf shooting off of a pinkish stem. Colors that work well together and happen to blend quite nicely with the color of bricks, between which spotted spurge can often be found growing, an art installation in the city.

Several Native tribes found use for it as a blood purifier and to treat urinary problems. The milky juice was used to wash cuts and rubbed on sores. The Cherokee people traditionally used a decoction of the plant to treat cancer. Many bold uses for such a humble plant. The plant bleeds milky sap when cut. The milky sap can cause skin irritation for some people, so do be careful when you are showing off your cool plant knowledge to a friend.

It is hated by farmers because it is toxic to livestock. But here in the city, we have nothing to fear. Spotted spurge fills in the cracks that we've left open, making use of what's available and nothing more. It serves as a food for insects who share this place and as a reminder to us that we have enough, we are enough, and we do enough.

St. John's Wort

Hypericum perforatum

A perennial of the Clusiaceae family, the Mangosteen family. A medium-sized family with 1,010 species that are bound together by their resinous sap (pun intended).

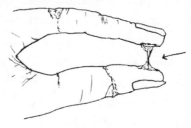

Habitat: Grows best in rich soil in full sun but also tolerates drought and compaction. Can be found in landscaped areas, public parks, rubble dumps, and median strips, as well as along waterways.

Came from Europe. Arrived here shortly after European colonization in the 1600s and was likely re-introduced a few times after that.

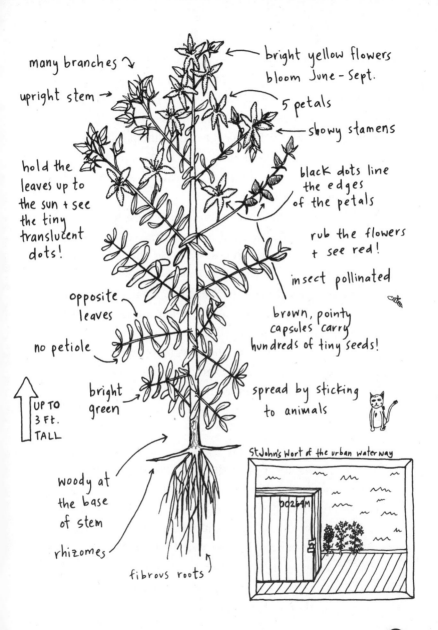

Each leaf is covered in tiny transparent dots that look like little holes or perforations when held up to the sun, which is where the species name *perforatum* comes from. The common name *wort* means "herb" in Old English. In its native lands, it has been used to treat ulcers, rheumatism, diarrhea, fevers, digestive disorders, bladder problems, pulmonary problems, and nervous disorders. Quite the list. A Russian proverb says, "It is impossible to make bread without flour as it is to heal people without St. John's wort."

St. John's wort didn't stop there; it was a known spiritual healer as well. In Celtic tradition, it is said to be very protective and was planted near the house or hung indoors to keep evil spirits at bay. In Germany, it was used to ward off the demons of melancholy, and its Yiddish name is *shedim* shuts, translating to "demon protection." When it was introduced to North America, some Native tribes added it to their medicine cabinet. They used it for fever and bowel movements and as a wash on infants to give them strength. The medical industry of North America still honors the -wort of this plant: St. John's wort has been popularized as a medicinal herb to treat depression.

St. John's wort was able to carve a home out of the city, and its prolific seeding ensures that it will be here for generations. A single plant can produce up to 100,000 seeds and those seeds can remain alive for up to ten years waiting to sprout. It is judged for crowding out native plants and is considered a noxious weed in nine states and twenty countries.

The striking yellow star-like flowers are a joy to happen upon. Have you ever seen them? They would be prized for their beauty if they weren't so common.

Stinging Nettles
Urtica dioica

A perennial of the Urticaceae family, the Nettle family. Some taxonomists link the Nettle family with the Moraceae family, the Mulberry family!

Habitat: Grows best in moist, fertile soil in either sun or shade. Can be found along waterways, at the edge of woodlands and in drainage ditches and community gardens

Comes from Eastern North America, Asia, Northern Africa, and Europe and is naturalized in temperate climates worldwide.

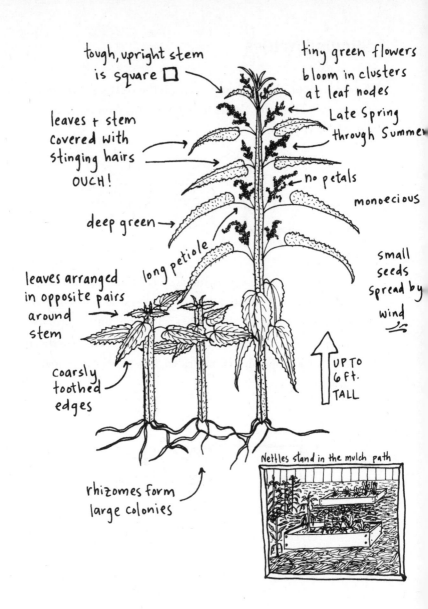

Nettle is derived from the Anglo-Saxon word for needle. In fact, many of the common names throughout the world reflect the plant's needle-like stinging hairs. Have you ever touched stinging nettle? It is deserving of its name: it stings and even more so if you're taken by surprise.

The sting wasn't enough to keep people away, however. Humans have long known and cherished nettles for their rich offerings. We have been eating nettles as a nutrient-rich green, using them as medicine to increase body immunity and overall vitality, weaving nettle fiber into cloth, ropes, nets, and twine and turning it into a green dye and paper for millenia. There is a long history of varied use by Native tribes across North America, and it was one of the Nine Sacred Herbs of the Anglo-Saxons. This plant was valued and honored. In Denmark, burial shrouds that date back to the Bronze age (3000–2000 BCE) were found made of nettle fiber.

We even called on the infamous sting to heal us. People whipped themselves (still do) with the stinging stems, in a process called urtification, to increase circulation as well as to relieve sciatic pain and arthritis. The sensation is similar to jumping into a freezing body of water; in the moment, it hurts, but once you sink in, the healing happens. The sting gratefully goes away when nettles are cooked as a potherb, in a soup, or as tea. And so nettles are easy to ingest to receive their benefits.

Visiting nettles is like visiting an old friend. I know in my bones the connection between nettles and the ancestors. There is such a rich history here in our relationship with nettles: let's bring back the love.

Virginia Pepperweed

Lepidium virginicum

A winter or summer annual of the Brassicacea family, the Mustard family. Related to Brussel sprouts, kohlrabi, radish, and watercress: mmmmm, fixins for a salad! And maca, known to increase male fertility.

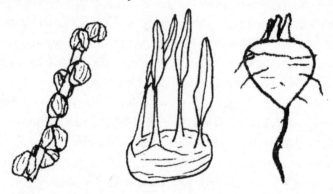

Habitat: Thrives in sunny, compacted soil. Can be found in public parks, landscaped areas, vacant lots, median strips, building foundations, sidewalk cracks, and rubble dumps.

Comes from Eastern North America.

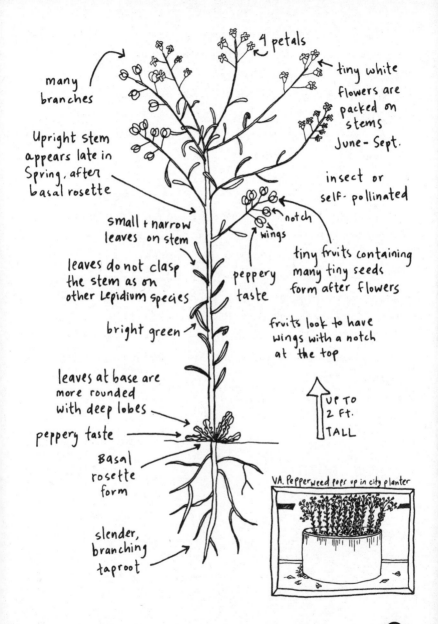

Virginia pepperweed is despised in agriculture because if cows eat the leaves, the milk produced afterward will be tainted with a peppery taste. Does that mean we have to hate it in the city too? Here, it fills in ground left bare by the movement of the city. The tiny white flowers and funky-shaped seed pods bring a softness that is juxtaposed against the hard structures of the city.

It was eaten as food in pre-colonial times, enjoyed for its peppery taste. Virginia pepperweed was also used medicinally as a poultice to counteract poison ivy and croup and to draw out blisters. The seed pods can be used like pepper, the source of another common name, poor man's pepper.

It seems we swapped members of the Brassicaceae family with Europe as Virginia pepperweed has naturalized there, while shepherd's purse has found a home here. Plants, like humans and other animals, travel the world looking for a suitable place to call home. Let us commend Virginia pepperweed for being able to thrive in city conditions and give thanks that it is still here with us.

Wild Carrot

Daucus carota

A perennial of the Apiaceae family, the Carrot or Parsley family! Related to the cultivated carrot that we know and love, as well as many other spices, including dill, carraway, coriander, fennel, lovage, parsley, and anise.

Habitat: Loves dry, sandy soils. Common in coastal meadows in its European habitat. Can be found on railroad tracks, rubble piles, vacant lots, landscaped areas, and building foundations.

Came from Eurasia and North Africa. Unknown how it traveled: perhaps by way of a broom?

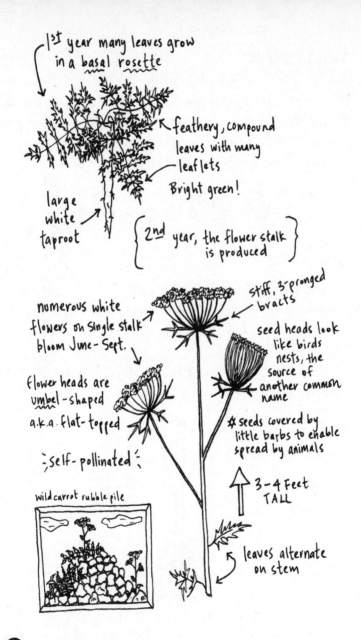

Wild carrot is the wild ancestor of cultivated carrots, which were reportedly first developed in Afghanistan. Wild carrot root is edible too, but not as super sweet as our beloved carrot. The seeds have a long history of use as a morning-after contraceptive in Europe and to reduce female fertility in India. Several Native tribes added the plant to their diet and healing regimens once it was introduced to this land.

Humans aren't the only beings to take advantage of wild carrot's spread. Wasps, bees, beetles, and butterflies across the US drink its nectar. Rabbits and deer eat the foliage, and European starlings use it as nesting material.

There is a mystical quality to wild carrot: some flower umbels have a single dark purple flower in the center of the white flowers known as the fairy seat. To add more color to the blooms, cut the flowering stalk and place it in colored water, the flower head will take on the color it is placed in! Red, blue, orange, purple—have a little botanical fun!

Wild carrot can turn a city street into an urban meadow, reminding us of the soil beneath the concrete. The flowering stalks can make us stop in our tracks and feel the earth under our feet, reconnecting us to our source. And what a blessing that is.

Yellow Rocket

Barbarea vulgaris

In the Northeast US, yellow rocket is a biennial of the Brassicacea family, the Mustard family. Related to these yummy annuals: cauliflower, broccoli, turnip, and kale.

Habitat: Prefers full sun in moist or dry conditions in nutrient-rich soil. Grows smaller if in poor, dry soil. Can tolerate some shade. Can be found in vacant lots, rubble dumps, landscaped areas, drainage ditches, highway banks, and riverbanks.

Came from Eurasia. Unknown how it traveled here. Perhaps by getting caught in a ship's floorboards?

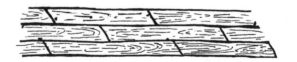

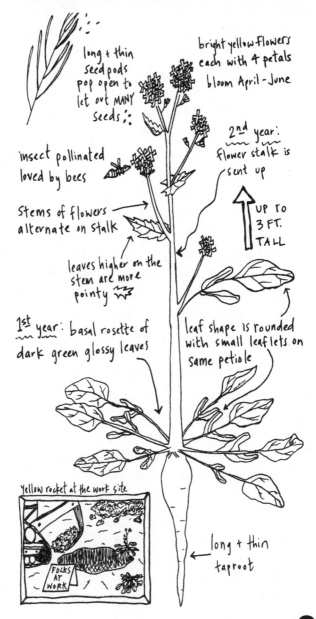

Yellow Rocket produces such bright yellow flowers that when they bloom en masse, it is a sight to behold. It is possible to spy such an event for, due to its prolific seeding, yellow rocket can take over large open areas—and large open areas are commonplace in a constantly shifting city.

Barbarea vulgaris, the botanical name, tells us two things. The first, that the greens emerged in Europe around December 4, the feast day of St. Barbara. The second, that it is very common, or was common, in Europe. It has now become common here as well and once you recognize yellow rocket, you will see it everywhere.

That is a good thing for your taste buds because if you choose to harvest, the early spring greens can be eaten as a potherb. The earlier the better: they get bitter as they age, similar to some people. (Just kidding.) The flower stalks can be eaten before the flowers open and taste similar to broccoli rabe. The Cherokee and Mohegan tribes enjoyed the flavor and adopted it as a food when it was introduced to this land.

Like all plants with yellow flowers, yellow rocket has a joyful, sunny disposition. It brings a smile to my face when I see it out on the block. Sometimes, when I need an extra boost, I harvest a few stalks for a wildflower bouquet, bringing their joy and my smile home with me.

Yellow Woodsorrel

Oxalis stricta

A summer annual of the Oxalidaceae family, the Woodsorrel family. The genus *Oxalis* contains over six hundred species!

Habitat: Thrives in nutrient-rich soil in full sun but tolerates both shade and drought. Can be found in vacant lots, sidewalk cracks, rubble dumps, median strips, landscaped areas, public parks, and drainage ditches.

Comes from Europe and North America.

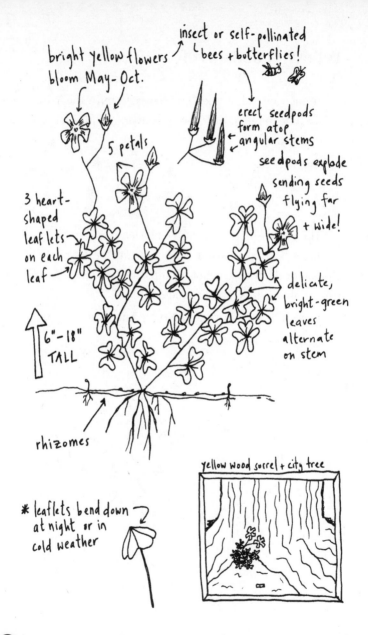

The genus name, *Oxalis*, is a Greek word for "sour" and the common name, sorrel, also refers to a plant with sour juice. Long before European contact, the Iroquois were using an infusion of yellow woodsorrel to "refresh the mouth" as well as for "summer complaints," and children of different tribes loved to eat the leaves for their fun taste. Yellow woodsorrel is well-known in both of its homelands for its sour taste.

The sweet and mostly sour taste comes from the presence of oxalic acid. Children still love to eat the leaves, which make a great snack while working in the garden. In fact, it can be beneficial to snack on yellow woodsorrel while you work. The Kiowa tribe's name for yellow woodsorrel means "salt-weed" and was traditionally chewed on long walks to relieve thirst. This shows evidence of an early understanding of salt-loss through perspiration and the ability of this plant to counteract it. Besides being edible, yellow woodsorrel can produce a yellow/orange dye when the whole plant is boiled.

It is impressive that this plant is able to survive in the city. Its native habitat is open woodlands and grasslands, ecosystems that are vastly different from the urban ecosystem. In the city, yellow woodsorrel still plays its part by feeding bees and birds. And it adds a distinct flavor to the neighborhood. Sitting pretty on the block, inviting us to add more fun and spice to our life.

Woody Plants

I am stating the obvious here, but woody plants are made of wood. Which is to say, their trunks, branches, twigs, and stems are covered in bark that is mainly made up of cellulose-walled cells encrusted by a substance called lignin. Life is definitely hard for woody plants in the city, but some still manage to make it to seventy-five years old. A tree might be your oldest neighbor on the block.

Trees bring more than we realize to our cities. People feel better and are actually kinder around trees. Hospital patients with trees in their sightline heal faster and feel less pain. Trees are physical medicine too. The bark, leaves, and fruit of trees have historically been used as medicine by people all over the world.

Trees ease the harsh conditions commonly found in cities, they
- Muffle urban noise
- Bring in birdsong
- Produce oxygen
- Remove carbon dioxide
- Cleanse air of pollutants
- Serve as windbreaks
- Serve as temperature regulators
- Fight soil erosion
- Conserve rainwater
- Reduce water runoff during storms

That's a lot of activity for a being that stands so still.

Perhaps most importantly, urban trees save lives. Heat is the leading weather-related cause of death in the United States. Urban centers tend to trap heat and have higher average temperatures than the surrounding land, a phenomenon known as urban heat island effect. We thought summer in the city was hot before, but now living in the time of global warming, our cities are getting even hotter. The temperature under a shade tree can be about twenty degrees less than the surrounding area. Trees keep us cool.

All of the woody plants discussed are deciduous, meaning they drop their leaves for winter. Visit each season to get to know your neighborhood trees in all of their different forms. See the branch form silhouetted against the sky in winter. Watch as the buds burst open in spring. Delight in the summer abundance that flourishes around you. Feel the light brush of the leaves as they fall in autumn—the trees, done with the last spectacular show of color, preparing for a well-deserved rest.

Black Cherry

Prunus serotina

A perennial of the Rosaceae family, the Rose family. A relative of delicious strawberry, apple, pear, apricot, cherry, almond, peach, and plum, as well as the ornamental and fragrant roses.

Habitat: Grows best in full sun and rich soil but tolerates drought and some shade. Can be found in public parks, along fence lines, waterways, and highways.

Comes from Eastern North America.

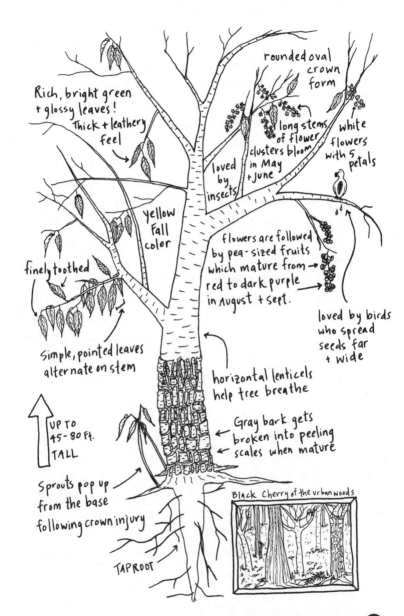

Black cherry is a pioneer species that depends on disturbance in wooded areas, its natural habitat. Thanks to the constant disturbance and tough living conditions of the city, the taller trees that would shade out black cherry and continue the process of ecosystem succession in the woods are not able to root down. Black cherry is part of the high biomass, high complexity climax community, and wildlife rejoices in what black cherry has to offer.

All kinds of bees and flies drink the flower nectar; ants drink extra nectar produced; caterpillars and beetles feed on foliage; songbirds, squirrels, raccoons, chipmunks, opossums, black bears, red fox, gray fox, and humans eat the berries. There is something for everyone, the value to wildlife is exceptional. Upon arrival to this land, Europeans found several Native tribes using the bark extensively for colds, coughs, and other ailments. Taking their lead, the arriving Europeans used the powdered inner bark to make wildly popular wild cherry cough drops. Different brands have reigned over the years, but the brand I always reach for is Luden's.

Black cherry enters our homes in other ways as well. An extract from its bark is used to flavor soft drinks, syrups, and candies, and the wood is highly sought after for furniture and cabinet making. Black cherry has naturalized in South America and is now considered invasive in several European countries after being widely planted in the Netherlands and parts of Central Europe in the early to mid-1900s. Seeds germinate in a variety of habitats, from sunnier to shadier and from drier to wetter, helping black cherry to spread its wealth.

Black cherry helps to reduce the intense heat of the city and feeds a wide range of animals in our neighborhoods. It grows and thrives with no input from us, a true gift.

Empress Tree

Paulownia tomentosa

A perennial of the Scrophulariaceae family, the Figwort family. A woody relative of common mullein.

Habitat: Grows best in full sun. Can tolerate drought. Can be found in vacant lots, building foundations, fence lines, and sidewalk cracks.

Came from temperate East Asia. Empress Tree was introduced to North America as an ornamental tree in 1844. It is speculated that its spread across the Eastern U.S. happened when seeds used as packing material to protect imported Chinese porcelain were discarded.

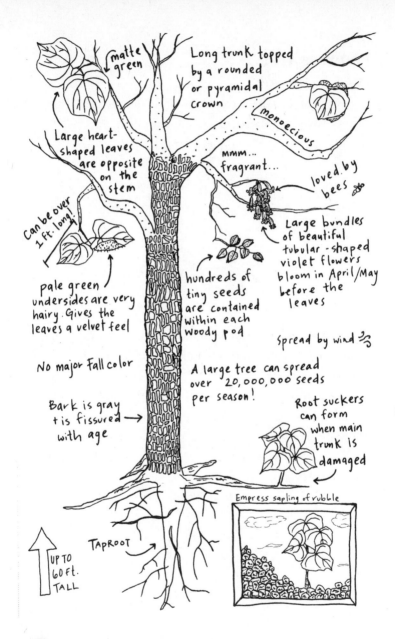

I don't pick favorites, but this tree is a beaut. I have seen it planted as an ornamental and pruned in a pollarded style that controls its height and, more importantly, lets us get a closer glimpse of the flowers. I have also seen it planted by the wind and growing in the seemingly smallest slivers of soil, a feat which never ceases to amaze me. Its prolific seeding ensures that you, too, can find it growing in a sliver of soil near you.

The empress tree has been cultivated on every continent in the world except Antarctica. The wood is prized in Japan and China for its light color, fine-grained texture, strength, and insect resistance. It is used to make everything from boxes and clogs to furniture and musical instruments, as well as for construction as posts and beams. Leaves can be cooked and eaten when other food is scarce or made into an infusion to use on skin ulcers or to prevent gray hairs.

An old tradition in Japan was to plant an empress tree at the birth of a daughter. When she was engaged to be married, the tree was cut down and used to make a dowry chest, furniture, and other wedding gifts. Such a beautiful way to mark the passing of time and major life moments.

Now, some empress tree enthusiasts want to plant them for phytoremediation efforts, to remove heavy metals from the soil. They also proclaim that empress trees can remove up to ten times more CO_2 than other trees and then release massive amounts of oxygen into the air, cleaning the earth from above and below.

The large, soft, pillowy leaves are open arms to a world of wonder. Let yourself fall freely into the plush beauty when the edges of the city are wearing you down.

Hackberry

Celtis occidentalis

A perennial of the Ulmaceae family, the Elm family. Related to classic city street trees American elm and zelkova, as well as slippery elm, whose bark offers a popular throat medicine.

Habitat: Can grow in a wide variety of habitats in sun or shade. Can be found in public parks and vacant lots, as well as along waterways, rock outcrops, and railroad tracks.

Comes from Eastern North America.

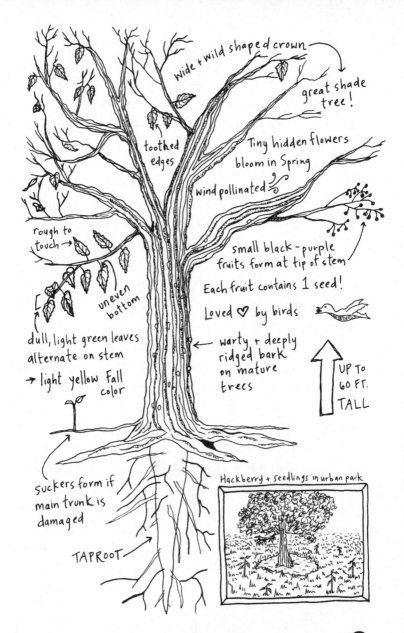

The native habitat of hackberry is limestone outcrops and moist bottomlands. On first glance that doesn't seem like it could easily translate to the urban environment. On closer inspection, we find that many buildings in cities are constructed from limestone, creating in the soil the same basic conditions that hackberry loves.

Basic soil is not a requirement however, hackberry can survive in a variety of soil types. Hackberry is tough: it can survive intense drought and is not susceptible to Dutch elm disease, the disease that ravaged urban populations of its cousin, American elm.

It is not much of a looker. In fact, it is often overlooked, especially in comparison to the grandeur of its cousins. Hackberry doesn't care; it still produces massive amounts of seeds that readily sprout on a variety of sites. Birds enjoy the fruits and in return for the sustenance, they spread seeds on their journey. The fruits are said to taste like raisins. I've never tried them but hope to harvest enough for a hearty snack one day. There is not much flesh around the pit, so plenty are needed to fill the tummy.

Ethnobotanist Enrique Salmon writes that Native Americans eat the berries fresh and dried, as well as use them as a spice. A decoction of Hackberry bark is used to regulate menses and to treat colds, sore throats, and venereal disease.

The hackberry emperor butterfly lays its eggs in the hackberry tree. The caterpillars love the leaves and feast until cocoon time. When the butterfly emerges it loves to stick around the hackberry tree, its wings blending right into the bark. Can you spot it?

The nipple galls are easy to spot. They look like warts on the leaves and are caused by tiny jumping lice; they are so common on hackberry that they can serve as an ID aid.

In the presence of a hackberry, the wise one tells us, don't mind how people see you: be true to yourself and the fruits of your being will flourish.

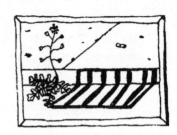

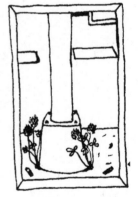

Norway Maple
Acer platanoides

A perennial of the Aceraceae family, the Maple family. Related to the beloved sugar maple of maple syrup fame, as well as classic ornamental Japanese maple trees.

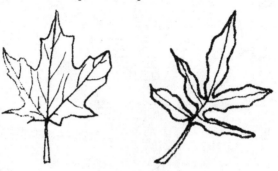

Habitat: Highly adaptable and can tolerate shade and compaction. Can be found in public parks, vacant lots, rubble dumps, waterways, and fence lines.

Came from Northern and Central Europe and Western Asia. Introduced by Philadelphian plant collectors in the mid-1700s.

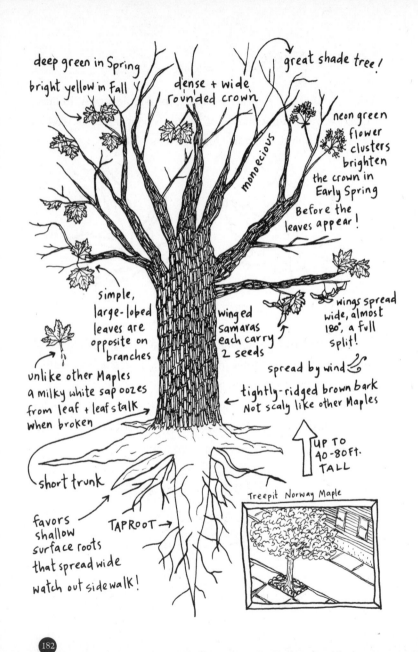

Norway maple is named for the top of its natural range, Southern Norway. In Europe, its native home, it is harvested for timber and planted extensively as a street tree, prized for the shade created by the dense canopy of large leaves. Upon its introduction to North America, many cultivars were bred, producing leaves of all shades of purple. In the 1950s and 1960s, after the urban American elm population was decimated by Dutch elm disease, Norway maple was planted to replace them. It is now one of the most commonly planted street trees in the Northeast.

It is a tree made for the city: fast growing and tolerant of wind, frost, pollution, salt, and a range of soil acidity. It is able to cope with pest and disease attacks better than other maples. Though, recently, the Asian longhorned beetle has been testing its coping capabilities.

Songbirds, squirrels, chipmunks, and rabbits feed on the seeds and spread them on their travels. This is where Norway maple gets a bad rep. Those seeds carried by animals and wind germinate in local forests and quickly shade out lower growing native plants. Conservationists become concerned because this changes the structure of forest habitats and reduces native diversity. Some states in the Midwest and Northeast list Norway maple as a noxious weed.

At least some animals of the forest find Norway maple palatable. Deer, elk, and moose eat twigs and leaves, bees drink flower nectar, and beavers feed on wood and bark. Maybe it's okay to let it be. At some point, we have to loosen our grip of control on the beings that share this world.

Quaking Aspen
Populus tremuloides

A perennial of the Salicaceae family, the Willow family. Related to the weeping willow tree and the Eastern cottonwood.

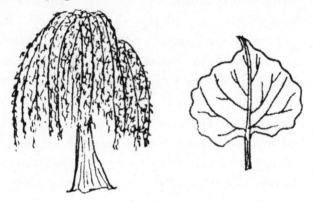

Habitat: Grows best in sunny, dry sites. Can be found in rubble dumps, railroad tracks, highways, parks, cemeteries, and beaches.

Comes from Northern, Central, and Eastern North America.

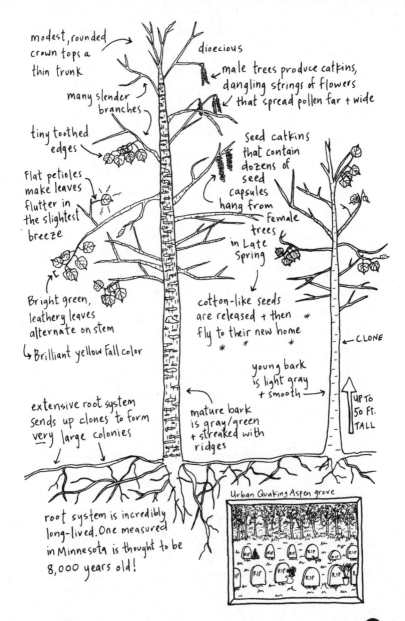

Urban Quaking Aspen grove

Quaking aspen is an important source of pulpwood for book and magazine paper. Are you holding the tree right now? If you love books, as I do, you will be happy to hear that quaking aspen is one of the most widespread trees in North America. It is a fast growing pioneer species that rises quickly in the full sun after a disturbance. Fire is its favorite, and quaking aspen plays its part in ecosystem succession by providing cover for shade-loving trees, such as maples and conifers, which will eventually grow to replace them.

Quaking aspen releases seeds with little hairs that fly in the wind, but it is the way it reproduces asexually by suckering that is truly fascinating. The extensive root system spreads out and sends up new tree sprouts that are clones of the first tree. An aspen grove in Utah is said to have about 47,000 stems; since they are all clones and the tree is dioecious, they would be either all male or all female. No wonder this tree is so widespread.

Many species of beings benefit from the offerings of quaking aspen. The yellow-bellied sapsucker drills holes into the bark to access the sweet sap; a wide variety of caterpillars and beetles eat the foliage; and beavers use the branches to construct their lodges and eat the inner bark for nourishment. People would also eat the inner bark in springtime. Its wood was used for fire and shelter; bark was both medicine and clothing material; and twigs turned into toys. Like other species in the willow family, the bark contains the compound salicin, the foundation of modern-day aspirin; it provides relief from pain, reduces fever, and is an anti-inflammatory.

The species name, *tremuloides*, comes from the Latin word *tremulus*, which means "quaking or trembling." The Onondagas named this tree *nut-ki-e*, meaning "noisy leaf." Both names honor the leaves when the wind blows. They look like shimmering ripples in the sky and produce the most beautiful sound. A sound that I cannot find words for.

Find one of these tress—can you hear it?

Staghorn Sumac

Rhus typhina

A perennial of the Anacardiaceae family, the Cashew family. Related to the delicious cashew, pistachio, and mango, as well as the ethereal smoke tree and the best-avoided poison ivy.

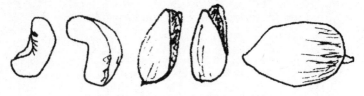

Habitat: Grows best in full sun and well-drained soil. Tolerant of road salt and compaction. Can be found in public parks, waterways, railroad tracks, vacant lots, fence lines, and rubble dumps.

Comes from Eastern North America.

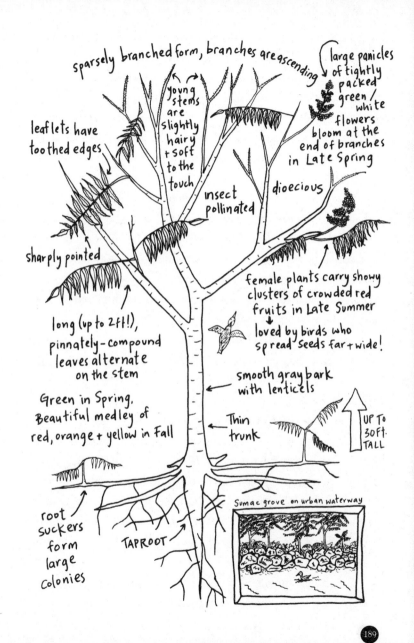

To see a staghorn sumac grove on a windy day is a spectacular sight, and in the fall with its brilliant display of colors? Fugghedaboutit! This North American tree whose native habitat is the woodland edge has found a way to survive in our ever-expanding cities, and our lives are richer for it. The staghorn sumac entered the city and continues to plant itself. It doesn't need our help, but we encourage its spread, planting it as a native option for erosion control on slopes.

Staghorn sumac sustains the diversity of wildlife that inhabits the city by providing food and shelter. Various bees, wasps, and flies visit the flowers for nectar. Berries are consumed by songbirds, and foliage is enjoyed by the caterpillars of many species of moths and some butterflies. Foliage, branches, and twigs are grazed by rabbits and deer—yes, there are rabbits and deer living in the city.

Staghorn sumac enjoyed a rich relationship with people of this land before colonization, one that continues to this day. The flowers, berries, bark, and root of staghorn sumac are used for various complaints, a dye is made from the berries and root, the young hollow stems are used as taps for maple sugaring, and berries are enjoyed in a summer tonic—soak berries in water, brew as a sun tea, and add a little sugar. Sumac lemonade! Yum, so refreshing in the summertime heat. The drink has a tart taste because of the ascorbic acid in the hairs of the berries. The berry hairs of a cousin, Sicilian sumac, are used as a spice in the Middle East, used to add tartness to a dish.

Staghorn sumac is named so because new growth is covered in velvet hairs, which mimics the fuzziness of a stag's horns as they begin to grow. Another common name is velvet sumac for the same reason. Whatever name you choose to call it, call it friend. Let it take us back to living in right relationship with the land.

Tree-of-Heaven

Ailanthus altissima

A perennial of the Simaroubacea family, the Quassia family. A small family (only one hundred species) of shrubs and trees that often have bitter bark.

Habitat: Grows well in full sun on dry, rocky, or sandy soil. Can tolerate shade, drought, salt, and pollution. Can be found along railroad tracks, fence lines, building foundations, vacant lots, and sidewalk cracks.

Came from Central China or Southeast Asia. Tree-of-heaven was introduced as an ornamental in 1784. Within fifty years of its introduction, it had become naturalized in both urban and rural areas.

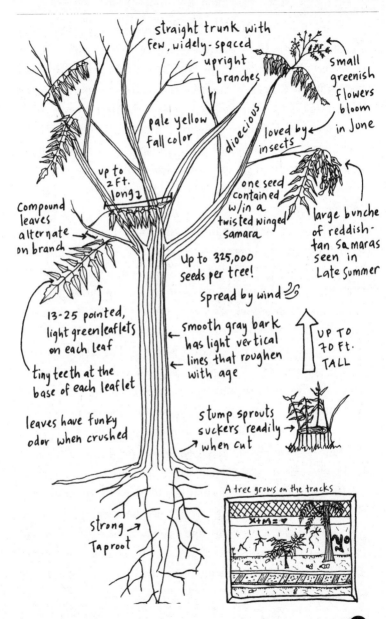

Born in Brooklyn, raised in Queens, my mom is a Brooklyn girl at heart. Fittingly, one of her all-time favorite books is *A Tree Grows in Brooklyn*. She, like Betty Smith, finds beauty in the everyday. And that is how the tree-of-heaven, or as mom calls it, *Ailanthus*, was one of the first trees I could name and why I've always regarded it with wonder. The large spreading crown provides an abundance of dappled shade that pours through the oh-so-tropical-looking leaves, at once transporting us to another land and keeping us cool in the summer heat.

Its fast growth and tall stature have been reflected in its name in many languages. *Ai* and *lanit* are the words for "tree" and "sky" in a language of the Moluccans from the Maluku Islands, words that were the inspiration for the genus name, *Ailanthus*. *Altissima*, the species name, is Latin for "very tall." Tree-of-heaven reaches for the sky. The larger the tree, the more seeds produced. Being that tree-of-heaven can reproduce sexually at just six weeks old and can live for about fifty years in the city, that's a lot of babies. No wonder that many US states list it as invasive and that it can be found on every continent except Antarctica.

Once germinated, tree-of-heaven produces a natural herbicide called ailenthene, which is shown to be toxic to thirty-five species of plants. This is extra assurance that would-be competitors would not be seeking the same place in the sky. City birds and insects don't seem to mind, happy to have tree-of-heaven serve as their food and habitat. Traditional Chinese medicine prescribes a tea of the bark for cases of diarrhea.

Even if we don't use it as food or medicine, it protects us from the sun's rays, muffles urban noise, cleans the air, and reduces water runoff. Tree-of-heaven also serves to open our eyes to the everyday beauty in our world. As Betty Smith so eloquently put it, "There's a tree that grows in Brooklyn, some people call it the Tree Of Heaven . . . it would be considered beautiful except that there are too many of it."

Virginia Creeper

Parthenocissus quinquefolia

A perennial of the Vitaceae family, the Grape family. Members of the grape family are famous in wines and juice and are eaten fresh or dried as raisins.

Habitat: Can tolerate a wide range of soil types and sun exposure. Best fall color in full sun. Can be found along chain link fences, brick walls, climbing trees in woodland edges, and scrambling along the ground.

Comes from Eastern North America.

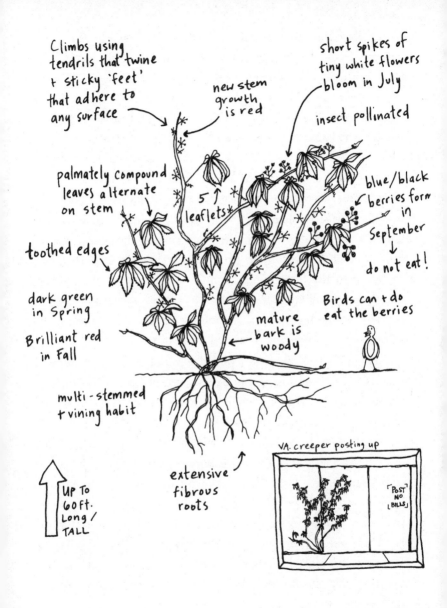

During the fall that followed my son's first blueberry picking season, I had a parenting lesson of a lingual matter. On a September walk around the neighborhood, my toddler son eagerly ran ahead and stopped to admire a chain link fence that was draped with Virginia creeper. By the time I got there, he had already eaten at least one berry and was happily showing me his harvesting skills. Oy! Yes, that is a blue berry but it is not the blueberry. What words to use to explain to a burgeoning talker the difference between these blue berries?

Nothing scary happened that day: he didn't seem to be bothered by the berries of Viriginia creeper, but they are best avoided. They contain tiny oxalate crystals that cause irritation to the lips, mouth, tongue, and throat if chewed. Songbirds do not experience this irritation and enjoy the berries for nourishment, including chickadees, robins, catbirds, downy woodpeckers, and many others.

The Kiowa people made a pink dye from the berries which was used on their skin and to dye feathers worn for the war dance. Virginia Creeper is commonly sold at nurseries for its fabulous fall foliage. It was exported to Europe, where it has since spread, as it easily escapes cultivation.

The species name, *quinquefolia*, refers to the five-pointed compound leaves. The five leaflets are a sure sign you are not looking at poison ivy: "Leaves of three, let it be. Leaves of five, let it thrive." Virginia creeper is a robust grower that has managed to remain despite the wave of human activity. It covers the naked walls and bare earth of the city with luscious deep green leaves that shine scarlet before they fall. A treat for us all.

White Mulberry

Morus alba

A perennial of the Moraceae family, the Mulberry family. A family that is often linked with the Nettle family.

Habitat: Grows best in full sun. Drought tolerant. Can be found in vacant lots, public parks, fence lines, sidewalk cracks, building foundations, and median strips.

Came from Eastern Asia. Introduced to North America from China in the 1600s during the craze to produce domestic silk. One Virginian had raised 70,000 white mulberries by 1664. Alas, the silkworm industry in North America fizzled and crashed, along with, I'm sure, many people's dreams and finances.

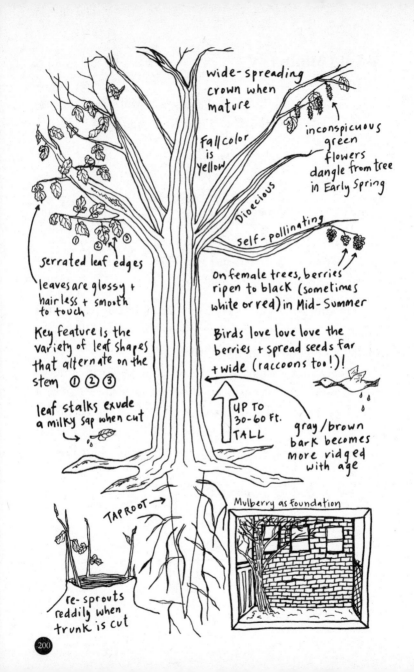

White mulberry is a whimsical tree, and what a history! How cool that one of humans' most prized fabrics was born of the blending of a caterpillar and a tree. The *Bombyx mori* silk moth caterpillars mix the sap of the white mulberry tree with their own chemicals and voila, silk! White mulberry silk enriched the Chinese Empire more than 2,000 years ago and was the reason for the fabled Silk Road across Asia, which quickened the meeting of Eastern and Western cultures.

In North America, white mulberry has found a home in our cities because of its tolerance to the extreme environmental conditions that mark city life. The native red mulberry cannot tolerate city life and so remains a tree of the woodlands. The white mulberry thrives in the city and lives to about seventy-five years old, a full life, if we let it.

A dear friend had a mature white mulberry growing in her Brooklyn backyard. Every summer, it was heavy with fruit, which she harvested to make mulberry jam! That is until, in her words, "It got totally butchered by our neighbors." Not everyone appreciates the berries, as they do leave a mess on the ground. And not everyone knows of their edibility. I sure didn't until witnessing my friend's resourcefulness—and I had been living alongside white mulberry my whole life.

The white mulberry is usually dioecious, having male and female flowers bloom on separate trees. Though it has been known to have both male and female flowers on one tree, and sometimes a white mulberry can bloom as a male one year and a female the next, totally throwing us for a loop!

The male flowers lack nectar to attract insects, so they produce massive amounts of pollen to be spread by wind, to the dismay of urban allergy sufferers. White mulberry at home in the city provides food and habitat for wildlife, helps with erosion control, and reduces the intensity of the urban heat island. It is one tree worth stopping to admire and marvel at for its pivotal role in human history.

Resources and Further Reading

BOOKS:

Plant Identification Guides:

Anderson, Wood Powell. *Perennial Weeds: Characteristics and Identification of Selected Herbaceous Species.* Iowa, Iowa State University Press, 1999

Cardina, John. *Lives of Weeds: Opportunism, Resistance, Folly.* New York: Cornell University Press, 2021

Del Tredici, Peter. *Wild Urban Plants of the Northeast: A Field Guide.* Ithaca: Cornell University Press, 2010

Elias, Thomas S. and Peter A. Dykeman. *Edible Wild Plants: A North American Field Guide to Over 200 Natural Foods.* New York: Sterling, 1982

Kaufman, Sylvan Ramsey and Kaufman, Wallace. *Invasive Plants: Guide to Identification and the Impacts and Control of Common North American Species.* Pennsylvania: Stackpole Books, 2012

Peterson, Roger Tory, and Margaret McKenny. *Peterson Field Guides: Wildflowers Northeastern/ North Central North America.* New York: Houghton Mifflin Company, 1968

Plotnik, Arthur. *The Urban Tree Book: An Uncommon Field Guide for City and Town.* New York: Three Rivers Press, 2000

Salmon, Enrique. *Iwigara: The Kinship of Plants and People.* Portland, Oregon: Timber Press, 2020

Sibley Barnard, Edward. *New York City Trees: A Field Guide for the Metropolitan Area.* New York: Columbia University Press, 2002

Zomlefer, Wendy B. *Guide to Flowering Plant Families.* Chapel Hill: The University of North Carolina Press, 1994

Herbal Remedies:

Bennet, Robin Rose. *The Gift of Healing Herbs: Plant Medicines and Home Remedies for a Vibrantly Healthy Life.* Berkeley: North Atlantic Books, 2014

Cohen, Debra and Adam Siegal. *Rediscovering the Herbal Traditions of Eastern European Jews.* California: North Atlantic Books, 2021

Duke, James A. *Handbook of Edible Weeds.* Florida: CRC Press, 1992

Soule, Deb. *The Woman's Handbook of Healing Herbs: A Guide to Natural Remedies.* New York: Skyhorse Publishing, 1995

Soule, Deb. *How to Move Like a Gardener: Planting and Preparing Medicines from Plants.* Maine: Under The Willow Press, 2013

Weiner, Michael A. *Earth Medicine-Earth Foods: Plant Remedies, Drugs and Natural Foods of the North American Indians.* New York: The Macmillan Company, 1972

Plant History, Lore, and Urban Storytelling:

Anderson, M. Kat. *Tending the Wild: Native American Knowledge and the Management of California's Natural Resources.* California: University of California Press, 2013

Chin, Ava. *Eating Wildly: Foraging for Life, Love and the Perfect Meal.* New York: Simon and Schuster, 2014

Dixon, Terrell F., Editor. *City Wilds: Essays and Stories about Urban Nature.* Athens: University of Georgia Press, 2002

Fromm, Erich. *The Art of Loving: Fiftieth Anniversary Edition.* New York: HarperCollins, 2019

Griffin, Susan. *The Eros of Everyday Life: Essays on Ecology, Gender and Society.* New York: Doubleday, 1995

Haupt, Lyanda Lynn. *Rooted: Life at the Crossroads of Science, Nature and Spirit.* New York: Little, Brown Spark, 2021

Kimmerer, Robin Wall. *Gathering Moss: A Natural and Cultural History of Mosses.* Oregon: Oregon State University Press, 2003

Kimmerer, Robin Wall. *Braiding Sweetgrass: Indigenous Wisdom, Scientific Knowledge, and the Teachings of Plants.* Minnesota: Milkweed Editions, 2013

Mabey, Richard. *Weeds: In Defense of Nature's Most Unloved Plants.* New York: HarperCollins Publishers, 2010

Ohlson, Kristin. *Sweet in Tooth and Claw: Stories of Generosity and Cooperation in the Natural World.* Ventura: Patagonia Works, 2022

Orion, Tao. *Beyond the War on Invasive Species: A Permaculture Approach to Ecosystem Restoration.* White River Junction: Chelsea Green Publishing, 2015

Richards, Gareth. *Weeds: The Beauty and Uses of 50 Vagabond Plants.* London: Welbeck Publishing Group, 2021

Sharkey, Erin, Editor. *A Darker Wilderness: Black Nature Writing from Soil to Stars.* Minnesota: Milkweed Editions, 2023

Smith, Betty. *A Tree Grows in Brooklyn.* New York: Harper Collins, 1943

Van Horn, Gavin. *The Way of Coyote: Shared Journeys in the Urban Wilds.* Chicago: The University of Chicago Press, 2018

WEBSITES:

American Botanical Council: herbalgram.org/BBC; bbc.com/future/article/20220217-the-strange-reason-migrating-birds-are-flocking-to-cities

Bio Tree: paulowniatrees.eu/learn-more/paulownia-environment/

Brooklyn Botanic Garden: bbg.org/gardening/weed_of_the_month

The Druids Garden: thedruidsgarden.com/2020/07/19/sacred-tree-profile-staghorn-sumac-rhus-typhina/

Eat the Weeds: eattheweeds.com/

Flowers of India: flowersofindia.net/

Gaia School of Healing: gaiaschoolofhealing.com/

Illinois Wildflowers: illinoiswildflowers.info/

Minnesota Wildflowers: minnesotawildflowers.info/

Missouri Botanical Garden: missouribotanicalgarden.org/

Native American Ethnobotany Database: naeb.brit.org

New York Times Coverage of the Blue Asiatic Dayflower: nytimes.com/2017/09/20/nyregion/the-true-blue-asiatic-dayflower.html

Plants for a Future: pfaf.org/

Poison Control: poison.org/articles/virginia-creeper-and-wisteria-toxicity-192#:~:text=The%20berries%20of%20the%20Virginia,mouth%2C%20tongue%2C%20and%20throat.

Uncultivated: uncultivated.info/

Glossary of Terms

Adventitious root – A root arising in an unexpected position grown from non-root tissue

Annual – A plant who lives its entire life cycle (seed to seed) in one growing season

Anther – The pollen producing part of the stamen

Basal rosette – A type of plant growth in which leaves form a circle at the base of the plant

Biennial – A plant who lives its entire life cycle (seed to seed) in two growing seasons

Biomimicry – An approach to innovation that emulates nature's time-tested patterns and strategies

Crown – The part of a woody tree or shrub that extends from the trunk or main stem (branches, stems, and reproductive structures)

Deciduous – Plants whose branches are bare of foliage for some part of the year

Dioecious – Bearing male and female flowers on separate plants

Evergreen – A plant, usually a woody perennial, bearing leaves throughout the year

Filament – The stalk of a stamen

Herbaceous – A plant with soft, green tissue (no wood) above the ground

Lenticels – Spongy areas in the cork of stems (and other plant parts) that allow gas exchange between the atmosphere and the internal tissues

Monoecious – Bearing separate male and female flowers on the same plant

Mycorrhizae – Symbiotic relationships that form between fungi and plants when fungi colonize the root system of a host plant

Ovary – The lower part of the pistil that contains the ovules

Panicle – A loose, branching cluster of flowers

Perennial – A plant living through several (or hundreds of) growing seasons

Petiole – The stalk of a leaf

Phytoremediation – The onsite use of plants to reduce contamination of soil or groundwater

Pistil – The ovule-bearing organ of the flower composed of a stigma, style, and ovary

Prostrate – Growth is trailing on the ground

Rhizobium bacteria - Aerobic bacteria that penetrate the root hair cells of legumes, thus establishing a symbiotic relationship. Legumes provide carbohydrates while the bacteria fixes nitrogen in the soil.

Rhizome – An underground, horizontal stem which puts out side shoots and adventitious roots

Root crown – The part of the roots from where the stem arises

Samara – A winged fruit containing a single seed

Stamen – The pollen bearing organ of the flower, composed of a filament and anther

Stigma – The top part of the pistil that is receptive to pollen

Style – The elongated section of the pistil between the ovary and stigma

Succulent – Fleshy, soft, and thick

Sucker – A vigorous, upright growing stem that grows from the base of the tree or from the root system

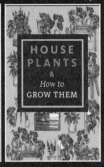

SUBSCRIBE!

For as little as $15/month, you can support a small, independent publisher and get every book that we publish—delivered to your doorstep!

www.Microcosm.Pub/BFF

Other books about Green Self-Empowerment: